International Conference on Design & Construction of LNG Ships 2016

London, United Kingdom
26 - 27 October 2016

ISBN: 978-1-5108-8297-3

Printed from e-media with permission by:

Curran Associates, Inc.
57 Morehouse Lane
Red Hook, NY 12571

CURRAN ASSOCIATES INC.
proceedings
.com

Some format issues inherent in the e-media version may also appear in this print version.

Printed by Curran Associates, Inc. (2019)

For permission requests, please contact The Royal Institution of Naval Architects
at the address below.

The Royal Institution of Naval Architects
8-9 Northumberland Street
London, WC2N 5DA
United Kingdom

Phone: 44 207 235 4622
Fax: 44 207 259 5912

publications@rina.org.uk

Additional copies of this publication are available from:

Curran Associates, Inc.
57 Morehouse Lane
Red Hook, NY 12571 USA
Phone: 845-758-0400
Fax: 845-758-2633
Email: curran@proceedings.com
Web: www.proceedings.com

CONTENTS

**Unavailable at time of going to press*

The Royal Institution of Naval Architects

INTERNATIONAL CONFERENCE

Design & Construction of LNG Ships

26-27 October 2016, London, UK

Papers published in these conference Proceedings are peer reviewed.

THE ROYAL INSTITUTION OF NAVAL ARCHITECTS
8-9 Northumberland Street
London WC2N 5DA

Telephone: 020 7235 4622
Fax: 020 7259 5912

LIQUEFIED NATURAL GAS SHIPPING HISTORY AND LNG SHIPPING IN THE 21ST CENTURY

A Clifton, Society of International Gas Tanker and Terminal Operators Ltd, UK

SUMMARY

As LNG shipping embarks on its second half century, what we should not forget is the very proud safety record we have as an industry. SIGTTO remains the industry leader for best practice and technical support for liquefied gas shipping and terminals and at the forefront of maintaining this safety record.

The LNG industry continues to expand and introduce new technologies. Larger ships with new types of propulsion system are now in service and the fleet continues to grow apace. FSRUs and FLNG vessels are also now part of the industry. All these advances ensure that there are many challenges in the liquefied natural gas shipping industry today.

This paper describes some examples of how this is reflected in the society's recent activities with respect to LNG vessel design and construction and also looks at the early history and designs of LNG vessels.

1. INTRODUCTION

LNG shipping's excellent safety record is the result of several factors, ranging from the underlying engineering through the operational procedures to the technical competence of operators. From the very first LNG vessels of 50 years ago to the first IGC code vessels right through to the vessels trading today, extra safety margins in safe design and operation have been factored into these vessels.

LNG is cryogenic; it is a liquid; and its vapours are flammable. It is not without its safety concerns and risks it, however, can be produced, transported and regassified as safely, and in most cases, more safely, than other liquid energy fuels. Public confidence in the safe transportation of LNG is essential, in particular, educating the public that the LNG shipping industry has an exceptional safety record.

2. THE EARLY HISTORY OF LNG SHIPPING

The pioneering cargo of LNG was carried across the Atlantic Ocean by the 5,000m3 Methane Pioneer at the start of a series of trial shipments in 1959.

For the development of the Methane Pioneer, Constock Liquid Methane Corp, a subsidiary of Constock International Methane, embarked on an ambitious research programme to verify the commercial feasibility of LNG transport by sea. The investigative work encompassed innovations in gas processing and liquefaction techniques, the evaluation of materials, ship designs, cargo-handling systems and storage tanks. Consultants from universities were employed on a part-time basis to translate the research results into design.

It was realised from the outset that the shipment of LNG by sea would pose special technological challenges. For a start, the design of the cargo tanks would be complicated by factors such as ship motions and the need to keep tanks firmly in position; independent expansion and contraction; ship hull deflections; and substantial temperature gradients during tank loading and discharging.

The ship chosen for the conversion was Normarti, a World War 2 Victory dry cargo ship of the C1 type, and the work was carried out at the Alabama Drydock & Shipbuilding yard in Mobile, Alabama. A dry cargo ship was chosen because it offered large double bottom, and wing tanks for the substantial amount of ballast the ship would be required to carry to achieve a suitable degree of hull immersion with the low density LNG cargo.

The result of the conversion was the 5,000m3 Methane Pioneer, a landmark vessel in the annals of LNG shipping. The ship was fitted with five prismatic tanks of aluminium, and balsa wood insulation was fitted to the ship's inner hull to a thickness of 0.3m. Balsa was the only material available at the time able to meet the required, stringent performance criteria. Besides providing good insulating properties, Methane Pioneer's insulation had to be able to perform as a secondary, liquid-tight barrier in case an aluminium cargo tank should fail. The balsa at the bottom of the tank had to bear the tank's laden weight and withstand the stresses induced by the ship's motions in a seaway. It also had to accommodate the thermal stresses associated with ambient temperature on one face and -162°C on the other without yielding. Finally, it had to be able to maintain its structural integrity in a fire situation for a period of four hours.

Methane Pioneer departed the Constock terminal at Lake Charles on the Calcasieu River with its first trial shipment on 28 January 1959 and arrived at Regent Oil's deepwater jetty on Canvey Island on 20 February after a trouble-free, 27-day voyage thus proving that LNG can be carried over long distances at sea.

By 1964 the first purpose-built LNG carriers, the 27,400m3 Methane Princess and Methane Progress, were in service under a 15-year gas purchase agreement signed by the UK and Algeria.

The newbuilding contract for Methane Princess was placed with the Vickers Armstrong shipyard at Barrow-in-Furness in northwest England while Harland & Wolff in Belfast, Northern Ireland won the order for Methane Progress. The vessels were sisterships, and Vickers Armstrong, as the lead yard, took responsibility for drawing up the working plans and placing material orders for the pair. Each ship was to cost £4.75 million to build.

The cargo containment system on the two ships was designed by Conch International Methane Ltd. The system was based on that utilised on Methane Pioneer and for which Conch held the design patents. JJ Henry, the New York based consulting naval architect firm, was closely involved with the hull design of the two vessels while Shell supervised their construction. Each ship was fitted with nine 5083 grade aluminium, free-standing cargo tanks, installed three to a hold in three holds. Each tank weighed about 130 tonnes and was lifted in as a complete unit. 'Keys' were fitted to the top of the tank to locate the unit and allowed for expansion and contraction.

Each tank had a full height centreline bulkhead with two separate crossflooding valves near the bottom. A single JC Carter submerged electric cargo pump, with a capacity of 200 m3/hour and a 82m head, was placed at bottom of each tank on the port side. This was the first marine application of submerged electric cargo pump technology.

Only one side of each tank had a filling connection. As a result it was necessary to have the cross-flooding valves open throughout loading and discharge operations but closed at sea, for stability reasons. The tanks were insulated primarily with prefabricated balsa wood panels faced with maple leaf plywood, which was impervious to LNG. On the tanks' vertical sides the panels were supplemented with glass fibre. The top of the tank was insulated with a loose-laid mineral wool material called Rocksil.

The insulation system was also designed to act as a secondary barrier in event of leakage from the primary aluminium tank. In modern parlance these would be described as IMO Type A tanks.
Methane Princess and Methane Progress operated successfully throughout their service lives. They were easy to operate and popular with their crews, although one or two problem areas came to light. The data-logger, for example, was never very reliable and needed frequent attention from the manufacturer's representative to keep it working.

Both vessels also suffered fatigue cracks in their inner hulls, leading to water ingress to the insulation. Repair techniques were developed for this problem, and accounts at the time claimed that the insulation properties of balsa wood did not seem to degrade when it became soaked.

A more frequent problem was cracks developing in the insulation system. These were indicated by areas of frost on the inner hull, or 'cold spots'. An in-service technique was developed involving drilling through the inner hull at the site of a cold spot and injecting a resin adhesive to seal the crack. This seemed to work well, but it did mean that there had to be a regular inspection routine for the inner hull to check for cold spots. All credit to those involved that it was only five years after the successful Methane Pioneer trial shipments and three years after the UK government approved the project, that the first two purpose-built LNG carriers should enter service. This start-to-finish timetable compares very favourably with all subsequent LNG projects!

After the Methane Princess and Progress were delivered further LNG vessels were constructed in the 1960s and LNG shipping had firmly established itself.

3. SIGTTO

When LNG commercial operations commenced in 1964, there was no industry body dedicated to liquefied gas during the formative years of the industry.

Through the years 1977/1978 a number of operators of LNG vessels in mutual correspondence had expressed interest in the establishment of some form of association in which LNG carrier operators internationally could assist each other in problems of safety and reliability and in the maintenance of high operational standards within their expanding industry. Stemming from this interest a series of meetings was convened through December 1978 to March 1979 in order to discuss the need for and possible formation of such an association. The first meeting took place at the Princess hotel in Hamilton, Bermuda on the 11th and 12th December 1978. The meeting had been called at the request of El Paso LNG company upon suggestions made by Compagnie Nationale Algerienne de Navigation (CNAN) and others that such an organisation could prove useful. At this time the intention was to call the new society "SIMTO", the Society for International Methane Tanker Operators".

Although the initial interest was centered upon LNG carrier operation, these meetings included representation from the broad field of liquefied gas transportation and it was quickly recognised that any such association would be the more valuable if it encompassed all forms of liquefied gas carrier operation (LPG and chemical gases as well as LNG) and if it also included the operation of terminals loading or receiving these cargoes.

In recognising this, it was appreciated that the problems attaching to safety and reliability were broadly similar throughout the various forms of liquefied gas marine transportation and that there would be great value in providing a channel of direct liaison between ship and terminal operators in the matters of safety and reliability at the ship/shore interface. It was seen that in developing and maintaining appropriate safety standards and in the public recognition in acceptance of these standards, the various components of the industry, LNG and LPG, ships and terminals were mutually supportive. The name SIGTTO was therefore decided upon as the most appropriate name for the new industry body.

In these formative discussions, the participants had the very ready help of senior executives of ICS and OCIMF both of whose bodies already included consideration of the liquefied gas industry within their activities and both of which already had consultative status with IMCO (now IMO). A most pertinent question clearly was whether the perceived aims and activities of the proposed new organisation could not be more appropriately achieved by the two existing organisations either singly or in combination. It was concluded, however, that an association directed exclusively to the liquefied gas marine transportation industry and providing under one umbrella a comprehensive forum for operators both on ship and shore would be the most appropriate means of achieving the desired objectives.

As a result of these formative meetings, the Society of International Gas Tankers and Terminal Operators Ltd came into existence as a Bermuda exempted company with limited liability and with availability of membership and shareholding in the company to those owning or operating a liquefied gas carrier or a liquefied gas marine terminal. The first board meeting was held in Bermuda on the 5th October 1979. The founder members were:

- EL Paso LNG Company
- Energy Transportation Corporation
- Malaysian International Shipping Corp
- BP Tanker company Ltd
- Moore, McCormack Bulk Transport Ltd
- P&O Bulk Shipping Ltd
- Gotaas Larsan Inc
- Marine Transport Lines
- Shell Tankers Ltd
- Exxon

A further board meeting with more members present was held in Houston on the 12th November 1979 At this board meeting Mr Maurice Holdsworth was appointed as the first general manager of the society and the founding President, Barry Hunsaker of El Paso Natural Gas. At this time there were 52 LNG vessels in service.

The first technical issues the society dealt with were as follows:

- Contingency planning
- Ship/shore linked emergency shut down
- Safe havens
- Cargo Strainers
- Training

Interesting to note that the SIGTTO technical committee of today are still discussing most of the above to some degree or another!

Within 1 year membership had grown to 30 members, including several terminal operators.

One of the immediate priorities was seeking the granting of consultative status at the IMO. This was achieved in 1982 when the membership had grown to 50 different organisations. By this time the society had an active GPC and a well-established Panel programme.

The society then progressed and steadily grew in membership to being acknowledged as the authoritative voice of the liquefied gas shipping and terminals industries. This position rests on its reputation the society quickly gained for impartiality and integrity in addressing operational and safety issues.

A number of publications were released to the industry as a result of issues addressed by the GPC and working groups.

In 2013 SIGTTO formed the Society for Gas as a Marine Fuel (SGMF), to oversee the use of LNG as a marine fuel.

The GPC and the Secretariat provide the vehicle through which the knowledge and information gathering within the organisation can best be promulgated to the members SIGTTO has published over 50 books, recommendations and guidelines – an average of two per year have been produced or updated.

The society is as strong now as it has ever been; the membership includes around 97% of the world's LNG vessels and terminals and around half of the LPG market. SIGTTO now has more members than ever before and remains the industry leader for best practice and technical support for liquefied gas shipping and terminals. SIGTTO has played a key part in achieving the industry's safety record.

The philosophy of the society is best described in the words of its founding President, Barry Hunsaker of El Paso Natural Gas:

"*We will best achieve our goals by sharing with each other our non-proprietary technical and safety information and operating experiences through open and frank discussion. Only in this way will each of us benefit from the experience and knowledge gained by all of us and thus maximise the safety of our operations. Remember, the industry will be judged by the record of its least safe operator. Let's help ourselves by helping that operator*".

4. LNG SHIPPING SAFETY

In 52 years of commercial operation, LNG vessels have carried over 85,000 cargoes, during this period there have been no loss of cargo tank containment and no onboard fatalities directly attributable to the cargo. This is a very impressive, in fact, unprecedented, safety record for the carriage of liquid hydrocarbons at sea in bulk.

This safety record down to several reasons which include, but are not limited to, a strong, overarching safety philosophy, robust design of equipment and systems, good operational & maintenance procedures, operating in excess of the minimum requirements and to best practice, high standards of training coupled with competency verification, the fact that the International Code for the Construction and Equipment of Ships Carrying Liquefied gases in Bulk (IGC Code) was developed based on actual experiences in the early days of LNG transport and an ability to share lessons learnt and to develop best practices as an industry.

Credit also needs to be given to the pioneers who's contributions to the early days of Liquefied gas shipping and also to the development of the IGC Code with its safety margins and safe design commitments, are also responsible for the excellent and unprecedented safety record. An example of the extra safety margins put into the initial design of the first LNG vessels in the early 1960s are the distance between the ships side shell and the cargo containment system. Long before double hulls were considered for oil tankers, LNG vessels had the surrounding ballast tanks to protect the cargo containment system from collision and grounding. This has directly contributed to the safety record that the industry enjoys today.

The safety record, as highlighted earlier, is extremely impressive. There have been incidents during the 50 year history, however, despite this, the safety record remains intact.

SIGTTO maintains a highly detailed database of incidents which have occurred onboard LNG vessels and within terminals. These incidents mainly concern machinery and cargo handling incidents. The machinery incidents include loss of propulsion or other major equipment failure and blackouts. Most of the cargo handling incidents have occurred at the ship/shore interface especially during connection and disconnection.

There have been three high speed groundings of LNG vessels. Despite substantial bottom damage in each of these incidents, none of these vessels suffered loss of cargo tank containment. There have also been other groundings involving slower speeds in port areas.

There have also been collisions involving LNG vessels, three in the last 18 months, one involving a large LNG vessel with a coaster off the Belgium coast and another large LNG vessel with a container ship in the Singapore Straits and the other involving a LPG vessel and LNG vessel in Tokyo Bay. The very first LNG carrier, Methane Princess, was struck by a vessel whilst alongside in Canvey Island, repairs to the side shell were required. Another LNG vessel was, almost unbelievably, struck by a submarine as it rose to periscope depth.

The common factor in these groundings and collisions is that there was no loss of cargo containment. This is a legacy of the extra safety margins and safe separation built into the design of these vessels by the pioneers of LNG vessel design which transformed into the IGC Code.

The LNG industry continues to expand, new technologies, new types of propulsion and larger ships than ever before are now in service. Floating LNG, regasification and reliquefaction are also now part of the industry. All of these activities ensure there are many challenges in the liquefied gas industry today.

Not least of these challenges is the supply of ship's crews, shore support staff and trainers to meet the additional numbers of trained and competent staff required for the increased fleet size. The SIGTTO competency standards for crews onboard both LNG and LPG vessels have become the industry best practice recommendation. This allows operators guidance as to the specific competencies each individual such possesses before serving in that rank. These standards are above and beyond the minimum STCW requirements. Similarly there is guidance available for terminal operators for their staff.

Educating the public is extremely important for liquefied gas shipping and the public needs to be made aware that Gas Carriers are not "Floating Bombs". Public perception is often that an incident on a gas carrier will result in a huge explosion that may harm people and property in the vicinity. The public needs to learn that these vessels are robust ships, soundly designed and constructed, well equipped with safety and emergency systems.

The public also needs to be aware that catastrophic events caused by hydrocarbon gases in the liquid phase are few, as an example, refrigerated liquefied gas tanks are highly unlikely to explode – but will burn until fuel is consumed.

Liquefied gas cargo handling procedures can be complex and the cargo itself is potentially hazardous. For these reasons, personnel operating gas carriers and gas berths require a thorough understanding of ship and shore equipment and cargo properties. They need to have available good operating procedures so as to avoid accident and emergency plans should be in place in case an accident does occur.

LNG is increasingly being carried as a cargo at sea in ISO containers on conventional container ships and the use of LNG as a marine fuel also brings with it new risk and fresh challenges.

5. REGULATIONS AND BEST PRACTICE

The mandatory The International Code for the Construction and Equipment of Ships Carrying Liquefied gases in Bulk (IGC Code) applies only to those vessels whose keel was laid on or after 1st July 1986.

The mandatory GC Code applies to vessels delivered after 30th June 1980; and the non-mandatory EGC Code applies to those vessels delivered on or before the 31st October 1976.

Amendments to the IGC and GC Codes introduced after vessels were delivered do not necessarily apply to such vessels.

SIGTTO recently facilitated the revision of the IGC Code on behalf of the IMO.

In 2008 SIGTTO formed 9 working groups under the auspices of a steering group, to carry out this work. Over a 2 year period, with nearly 140 experts, representing over 40 entities and 20 countries, SIGTTO ensured that this revision was delivered to IMO. The Code was approved at MSC 92 (May 2014). After many years, the revised IGC code was finally adopted at MSC 93 (June 2013) and entered into force on the 1st January 2016 and was implemented on the 1st July 2016. The revision is not retroactively applied and only applies to vessels contracted/keel laid after 1 July 2016.

It is interesting to note that the IGC Code is one of the few IMO instruments not to be drawn up following a major incident, SOLAS (Titanic), MARPOL (Torrey Canyon) and ISM (Herald of Free Enterprise) are some examples, this gives the advantage of the Code not being drawn into any kneejerk requirements and being able to develop the Code based on actual experiences in the early days of LNG transport.

With these challenges in the liquefied gas industry today and the advent of new players into the market there remains an ever present need for a supply of best practice and guidance to the industry. SIGTTO has noticed that Marine regulations are becoming increasingly focused on human factors which is forcing industry to take account of this important topic. Organisations that actively take steps now to address human factors issues will be best placed to succeed.

SIGTTO is addressing human factor issues through the Human Element committee which is looking at gas specific issues arising from incidents and near misses and keeping these issues on the agenda until they have been successfully resolved. The use of best practice in operations cannot be emphasized enough as there is little doubt that the LNG industry owes its safety record, in part, to the use of best practice.

6. LNG SHIPPING TODAY

Since the early days of LNG transportation, there have been many changes to the industry. LNG vessel size has risen considerably since the end of the last century. From a standard size of 125,000m3 and then 135,000 m3 the average size of vessels ordered rose up to the 160000 and 170000 m3 range. The Qatar "Q flex" and "Q max" vessels are by far the largest vessels trading today at 215000m3 and 265000m3 respectively.

The LNG fleet in service passed four hundred vessels in April 2014. With over one hundred vessels on order the fleet should pass five hundred in 2017. This is a remarkable achievement and is all the more remarkable that only as recently as 1997 did the fleet pass one hundred vessels.

LNG carriers are also amongst the most durable of all ships. Several have successfully traded well beyond their 40th year. In 2015 approximately 10% of the LNG fleet was in excess of 30 years of age. Some older vessels have been converted into floating regas and even floating liquefaction vessels. Some of these vessels could, conceivably, still be in operation beyond their 50th year.

Until recently LNG was a very conservative "club" with a small core of members however in recent years many new players have entered the industry. LNG is also being traded globally now, not regionally as in the past. There are several portfolio players with access to gas from multiple locations which require the vessels to be flexible and compatible with all terminals. However around 60% of LNG shipping remains on long term project work as there is a high capital investment in ships which often requires a long term time charter party to secure vessel financing, the daily operating costs of LNG vessels higher than most other parts of shipping.

In the first forty years of LNG shipping LNG carriers saw very few major step changes in vessel design and operations since they first came into service over. In the last twelve years, however, there have been some major changes with regard to propulsion options, vessel size, technology and trading patterns. LNG vessels were traditionally built with a reliable, but not very efficient, steam-turbine propulsion system, as boilers were the only means of consuming boil-off gas. Boil-off gas was generally viewed as waste gas that would be used as fuel. Within the last ten years, the introduction of Dual Fuel Diesel Electric (DFDE) propulsion systems (ships able to burn both diesel oil and gas) has created a jump in vessel efficiency. In real terms, a 145,000m3 steam vessel may have a total energy consumption, at the design speed and draught, of around 180 tonnes per day, whereas a DFDE equivalent would only need approximately 130 tonnes, representing an improvement of around 25-30%. Matching generator sizes so they operate at their maximum efficiency will further reduce the vessel fuel consumption and can be very beneficial for operating at a range of speeds. A DFDE operator will have a gas combustion unit and/or a reliquification plant installed for boil off gas management.

In the last two years vessels have been ordered and delivered using two stroke technology, as for DFDE vessels, these vessels will have a gas combustion unit and/or a reliquification plant installed for boil off gas management. The two stroke engines can be either high pressure or low pressure gas burning.

The Qatar "Q max" and "Q flex" vessels plus a handful of other vessels utilise conventional slow speed diesel engines without the option for gas burning in the main engine. These vessel must have both a gas combustion unit and a reliquification plant installed for boil off gas management. Interestingly, one of these vessels was recently converted to a high pressure two stroke MEGI engine.

In reducing the fuel consumption in the main engine the amount of natural boil off became too large resulting in wastage. Cargo containment systems have also seen a drive on efficiency with the 0.25% daily boil off gas from the 1970s being reduced to 0.15% in 2000 and now vessels are being delivered capable of boil off rates of 0.08% or even less. This ensures that less LNG boils off on passage, less in consumed in the main engine and more is discharged. Shipyards have traditional optimised the hull form around the ship build contractual guarantee value at 19.5 knots when the vessel is laden. As a consequence, the yards compete at this figure and designs are evaluated at this point. However the 19.5 knots figure is a legacy from liner trades and when vessels had higher boil off rates. It is now the minimum standard to which nearly all LNG carriers are built as it is the figure in the industry-standard Time Charter Party (TCP) agreement for LNG vessels (ShellLNGTime1). Independent LNG carrier owners typically specify a higher design speed (19.9 knots for example), partly to ensure that they have enough margin to meet TCP requirements, especially when carrying out maintenance on one generator engine at sea.

SIGTTO has produced guidance regarding all technical aspects of an LNG vessel transit through the new locks in the Panama Canal, this publication is titled "Guidance for LNG Carriers Transiting the Panama Canal" and has been available since June 2014.

The publication also contains recommendations on safety, minimum equipment levels and good operating practices. Exceptional support and co-operation from the Panama Canal Authority in compiling this document. Many existing LNG vessels plus many of those under construction will require a degree of modification in order to transit the canal. Most of these modifications are expected to be of a minor nature however the publication highlights in detail precisely what will be required. One major issue is that wires will not be permitted in the canal meaning all LNG vessels will have to have non-wire moorings on drums in order to transit. To have to remove wires from drums and replace them again after transit is not very sustainable for vessels on regular runs through the canal.

7. THE FUTURE

LNG vessels of the 21st century are more efficient, complex and larger than the vessels of the past. There is likely to be large scale demand for the bulk transportation of natural gas in its liquefied form for many decades to come. New projects will require new vessels and a lot of the older ships will be replaced in due course.

The biggest impact of the new locks at the Panama Canal for LNG vessels is likely to be in determining maximum LNG vessel size for future new builds. The 49m beam will be the limiting factor and hip-owners will be wary of ordering a vessel in excess of this as this will limit future trading options and charter opportunities. This will result in vessels of 175000-180000m3 being both the standard size and the maximum size to be built in the future.

The drive to improve efficiency will continue with it becoming increasing important to reduce boil-off rates of the cargo insulation system and to enhance operating efficiencies, such as fitting reliquefaction plants to retain the boil-off that would otherwise be wasted by sending it to the gas combustion unit.

There will also be a continued drive to improve the efficiency of the propulsion systems and to achieve optimum gas burning capability with reduced maintenance and down time. The main engine technology being developed for LNG as a marine fuel onboard conventional merchant vessels should assist this drive.

There is also likely to be a reduction in design speed, the 19.5 knots charter party speed, as mentioned earlier, is not so relevant today as boil off rates have vastly improved, a design speed of 16.5 knots would be more relevant and would also increase the efficiency of minimising the natural boil off, the amount consumed in the main engine and increasing the quantities discharged.

There is likely to be more specialist vessels ordered in the future such as FSRUs and FLNG units and, quite likely, a large scale unit with the ability to have LNG storage, regasification and power generation. This would have the capability to be brought into a location and provide direct power supply with minimal shore infrastructure required. Very suitable for areas hit by natural disasters or major engineering projects in isolated locations. The increased use of LNG as a marine fuel onboard conventional merchant vessels will see more small scale LNG vessels in operation and vessels dedicated as LNG bunker vessels.

8. CONCLUSIONS

A robust safety regime has been established and it is incumbent upon the shipping industry to make appropriate use of it, and where necessary, adapt it to suit particular circumstances, such as the use of LNG as a marine fuel, in order to preserve the exceptional safety record the LNG shipping industry currently enjoys.

We look forward to this very responsible industry continuing to ensure the safe transportation of liquefied gases by sea. SIGTTO remains committed to the safe design and operation of LNG shipping.

9. AUTHORS BIOGRAPHY

Andrew Clifton is the General Manager and Chief Operating Officer of SIGTTO (The Society of International Gas Tanker & Terminal Operators Ltd), he has over 30 years experience in the liquefied gas shipping industry. This includes 19 years at sea, three years as master, mainly on liquefied gas carriers, a first class honours degree in shipping operations and shore managerial experience over the last 12 years. The latter includes three years working at the UK's Marine Accident Investigation Branch (MAIB), marine superintendent with Golar LNG, almost two years in the SIGTTO Secretariat as a Technical Adviser and, most recently, over five years in Indonesia as LNG shipping operations manager for the BP Tangguh LNG project.

In this most recent posting Andrew was responsible for the project's shipping activities and overseeing their development, from the early construction days at Tangguh through to the commissioning and entry into service of the marine facilities and the ships built to serve the project.

Andrew was also the SIGTTO Panel chairman from October 2008 until becoming GM.

Andrew Clifton is the eighth SIGTTO General Manager. He is also the youngest General Manager and the first General Manager who was formerly a Technical Adviser at the Society.

FLOATING LNG – DESIGN DRIVERS AND TECHNICAL CHALLENGES

K W Hutchinson and **D C Dobson,** Babcock Energy and Marine Technology, UK

SUMMARY

Drawing on nearly three decades of combined experience in the design and evaluation of numerous new-build designs and conversion solutions for offshore Floating Liquefied Natural Gas (FLNG) vessel assets, the authors consider the history, development and imminent deployment of FLNG. In addition often conflicting and diverse design and operational criteria that have to be addressed coherently within the design process to facilitate the generation of robust and safe FLNG solutions are discussed.

In order to 'set the scene', the history and development of FLNG over the past few decades is discussed, including possible areas of operation / deployment and current vessels under construction and proposed.

The paper then discusses the regulatory framework and technical challenges that are faced by Naval Architects and Marine Systems Engineers in the development of robust, safe, compliant and near-optimal novel FLNG design solutions. This addresses both production and operational performance philosophies and criteria together with the management of environmental impacts, specifically with respect to principal dimensions, hull form, topology / layouts, safety, stability, motions, mooring, structural configurations and scantlings, prime movers, marine systems, electrical philosophy etc.

The paper also briefly highlights some possible future emergent technologies which may be deployed in support of the development of currently undeveloped hydrocarbon reserves

1. INTRODUCTION

Stranded subsea hydrocarbon reserves are those which cannot be economically exploited due to a number of reasons such as the field's composition, distance to market, extreme water depth, difficult subsea terrain etc. A number of technology options are available to the developer for exploiting the energy potential of such offshore reserves. One of the most popular over the past four decades is that of the Floating Production, Storage and Offloading (FPSO) vessel. In the simplest of terms, all FPSO facilities, or assets, can be thought of as consisting of two major components, the Hull and the Topsides.

1.1 HULL

The purpose of the Hull, which is sometimes referred to as the 'Barge' or in some extreme cases the floating 'Substructure', itself can be broken down in to serving three broad functions:

- providing the required footprint and support, both in terms of floatation and structure, considering both the dry and operational weights, the Topsides process modules and facilities; together with
- providing the required storage and offloading capabilities for the produced hydrocarbon liquids; together with
- providing the required accommodation and control spaces, marine systems etc. and quite often also elements of utility systems required to support the operation of the Topsides.

Hence, as the Hull provides the buoyant and stable facility platform it therefore, together with the environment in which the asset is deployed, dictates the requirements for the mooring system etc. It should be noted that if the function of accommodating the Topsides modules is not required then such a facility is termed a Floating, Storage and Offloading (FSO) vessel – such vessels are commonly associated with Fixed Platforms that require on-site storage of produced hydrocarbons, typically to date oil, which are then periodically exported to Shuttle Tankers for trans-shipment to market.

1.2 TOPSIDES

The Topsides consists of the process plant and utilities, typically arranged in separate modules. The utilities are those systems required to support the operation of the process plant, and include for example the main power generation which is usually dedicated but forms part of the overall vessel electrical infrastructure. Topsides also interface directly and significantly with the production risers and umbilicals, which connect the floating production facility with the subsea wellheads and manifolds, as these are either suspended directly from the sides of the Hull, if Spread Moored, or through the Turret, if such a mooring system is employed.

This paper will concentrate on briefly discussing the design aspects of a specific type of FPSO, namely a Floating Liquefied Natural Gas (FLNG) vessel facility i.e. one which is designed to process, liquefy and store, mainly, Liquefied Natural Gas (LNG) and subsequently offload it to LNG Carriers (LNGCs). A FLNG typically involves the transfer of mature previously exclusively land based natural gas liquefaction technologies to an offshore marine environment. In essence the hull design process of a FLNG is not too dissimilar to that for oil FPSOs. Basically, the design of an oil FPSO is in itself not too dissimilar to that of a large Oil Tanker, apart from that the local environment in which the FPSO is moored must be considered and that there is no need to consider propulsion, but obviously heading control via thrusters may be required - indeed, in the early days of FPSOs quite a number of relatively large Oil Tankers were converted to FPSOs for both benign and harsh environments. So whilst the hull design procedure for a FLNG can therefore be likened to that of a LNGC there are some subtle but not unsubstantial differences. Therefore, it is the intention of this paper to briefly highlight the key features and present some potential solutions related to the hull design of FLNGs.

2. FLOATING LIQUEFIED NATURAL GAS

The floating LNG supply chain basically consists of three elements:

- A FLNG offshore moored over the gas field;
- A LNGC for trans-shipment of the LNG to market;
- Either: a standard LNG reception Terminal onshore; or a Floating, Storage and Regasification Unit (FSRU) vessel moored near-shore.

Therefore, if we assume that both ends of the chain are floating assets then we have two broad vessel types, namely a FLNG and a FSRU

2.1 FLNG

A FLNG vessel is a facility which is permanently moored over typically an offshore, either lean or rich, gas field usually via a turret, but dependent on location a spread mooring system can be utilised. Typically, it receives the multi-phase well fluids via flexible risers, but the risers can be steel catenaries if the development philosophy allows. As with a (standard) oil FPSO, a FLNG separates the multi-phase well fluids and stabilises any associated condensate etc.

Note that there are alternatives to 'full FLNGs' i.e. facilities that are designed just to receive pure processed gas from, typically the shore, and hence tend to be moored near-shore or on jetties, as with FSRUs. These are however out with the scope of this paper, but it is worth noting the Topsides (liquefaction, power generation etc. only) can be significantly less than that of a 'full FLNG'.

Following separation, the gas is cleaned etc., as in a normal FPSO process train. However, instead of being compressed and, typically, exported via pipeline the gas is liquefied to form LNG, which is then stored in cryogenic tanks based on an appropriate Containment System (CS) commensurate with the metocean environment etc. If heavier fractions are present then Liquid Petroleum Gas (LPG) may also be produced as could Condensate, both of which would require dedicated storage facilities on the FLNG. The LNG, and possibly LPG and Condensate, are offloaded at the required intervals to suitable trading LNGCs, LPGCs or Shuttle Tankers etc. using an appropriate offloading system. These export tankers then transport the LNG, or other, cargo to market.

If the exporting LNGCs are required to berth alongside the FLNG then the design may allow for maintaining draught throughout the offloading evolutions. Typically, due to the large liquefaction trains, associated process equipment and large amount of power generation required, it can be the case that a FLNG design may be driven more by deck area requirements rather than storage capacity etc.

Typically, FLNGs are new-build Hulls designed for the specific process Topsides and environment, however for certain prospects conversions can be appropriate. The technology for hull design, being akin to that for LNGCs and large Oil Tankers and of course oil FPSOs, is proven and most of the major classification societies who are members of the International Association of Classification Societies (IACS) have rule sets and approval procedures to cover the specifics of FLNG hull design and safety. Hence, considering this and the fact that liquefaction technologies are mature on land, then FLNG presents a low technologically and environmentally risk solution for both operators and licensors i.e. governments.

Application of FLNG technology to stranded gas reserves has been proposed for locations worldwide, from those relatively close offshore to distant offshore fields, in shallow and extremely deep waters, in benign and harsh environments etc. Examples of the sea areas for which FLNG technologies have been proposed and solutions developed include the Arctic, off North America, in the Gulf of Mexico, off South America, in Europe, off both East and West Africa, in South East Asia and Australasia etc.

Obviously not all FLNG projects are in the public domain but those that are at the time of writing have been summarised in Table 1 by region in terms of those in-service, under construction and in various stages of design or planning, broken down in to new-buildings and conversions.

Description	Total	Africa	Australia	Brazil	Canada	Caribbean	China	Gulf of Mexico	Mediterranean	North East	Pacific	South America	Southeast Asia	Southwest Asia	TBD
In Service	**1**												1		
New Build	1												1		
Conversion	0														
On Order	**7**	2	1										1		3
Allocated	**4**	2	1										1		
Speculative	**3**														3
New Build	**4**		1										1		2
Conversion	**3**	2													1
In Planning	**38**	5	9	2	8		1	4	1	1		1	4	1	1
Prospective	**4**	1	1					1						1	

Table 1: Status of FLNG projects worldwide

Status	Construction	Name	Commissioning	Location	Region	Water Depth (m)	Mooring	LNG Production (mmcf/d)	LNG Storage m^3	Condensate Production (bbl/d)	Condensate Storage (bbl)
On Station	New-build	PFLNG 1	2016	Malaysia	SE Asia	200	Ext. Turret	1	177,000	-	120,000
On Order	Conversion	Golar FLNG 1	2017	Cameroon	Africa	-	Spread	2	125,000	-	-
On Order	Conversion	Golar FLNG 3	2020	Equatorial Guinea	Africa	1,800	Ext. Turret	2	126,000	-	-
On Order	New-build	Prelude	2017	Australia	Aus / NZ	250	Int. Turret	4	220,000	35,000	1,300,000
On Order	New-build	PFLNG 2	2020	Malaysia	SE Asia	1,150	Ext. Turret	2	177,000	-	-
On Order	New-build	Caribbean FLNG	2016	TBD	TBD	100	Jetty Moored	1	16,100	-	-
On Order	New-build	Exmar Speculative	2018	TBD	TBD	50	Jetty Moored	1	20,000	-	-
On Order	Conversion	Golar FLNG 2	2019	TBD	TBD	-	-	2	125,000	-	-

Table 2: Details of completed FLNGs and those ordered

From Table 1 it can be seen that currently 42 projects are at various states from initial planning to engineering design. The first FLNG vessel was completed during 2016, a further 4 new-build FLNGs are currently under construction with another 3 being converted from existing LNGC tonnage. From Table 2, which details the current and contracted FLNG fleet, it can be seen that LNG production rates range from approximately 4 mtpa down to 1 mtpa with associated storage from approximately 220,000 metres^3 to 16,000 metres^3, and various mooring systems are utilised due to differing locations and water depths. It is also interesting to note that at present 3 of these FLNGs are speculative in nature and currently 1 is suspended.

2.2 FRSU

A FSRU vessel is a facility which receives LNG offloaded via, typically, hard-arms from a LNGC, stores and then, as required, regasifies the LNG and exports the natural gas to shore via a subsea pipeline. Because of the requirement to allow LNGCs to berth alongside and in order to service a location or specific facility onshore, such as a gas powered power station, a FSRU is typically moored in a sheltered inshore location via a buoy or directly at a jetty, similar to those at LNG terminals.

FSRUs can be either new-build, dedicated or a LNGC designed to be utilised as a FSRU, or a conversion from older conventional LNGC tonnage employing both Self-supporting Spherical Type 'B' and Membrane CSs. Hence, the technology for hull design is that for LNGCs and as such is proven and covered by the rule sets of the major IACS classification societies. Therefore, considering this and the fact that liquefaction technologies are purely transfer of mature but previously exclusively land based technologies to a near-shore marine environment and are neither heavy or of a large footprint, then a FSRU presents a technologically and environmentally low risk solution for both operators and licensors i.e. Flag States.

An FSRU is a flexible, cost-effective way to receive and process shipments of LNG where no onshore infrastructure exists or gas requirements are small or temporary in nature. In addition, FSRUs are cheaper and quicker to build than onshore plants, hence are a suitable technology for low commodity etc. projects. Inherent in the FRSU concept is the ability for redeployed globally to meet market demand.

Table 3 below gives a breakdown of the FRSU fleet by region in terms of those vessels currently in-service, under construction / conversion and in various stages of design or planning, broken down in to new-buildings and conversions.

Description	Total	Africa	Australia	Brazil	Canada	Caribbean	China	Gulf of Mexico	Mediterranean	North East	Pacific	South America	Southeast Asia	Southwest Asia	TBD
In Service	**19**	1		3			1		1	1		3	2	7	
New Build	11	1		1						1		3	1	4	
Conversion	8			2			1		1				1	3	
On Order	**12**					1	6			1		2	2		
Allocated	6					1				1		2	2		
Speculative	6						6								
New Build	10						6			1		2		1	
Conversion	2					1								1	
In Planning	**41**	8		2			5		3	1	1	3	9	9	
Prospective	**6**						1		2			1	1	1	

Table 3: Status of FSRU projects worldwide

It can be seen that currently there are 19 FSRUs in service worldwide, 8 of which are converted from existing LNGC tonnage, and 10 are under construction and 2 being converted. It can also be seen that currently 47 projects are at various states from initial planning to engineering design.

2.3 LNG MARKET

At present the majority of new LNG projects, and hence increase in liquefaction capacity, are planned or under construction in Australia and North America. As we all know the current climate for offshore assets is difficult as, in response to low oil prices, investment in most future projects has been either delayed or cancelled.

It is generally agreed that there will be excess LNG supply up until 2019, and hence as a result no LNG export projects, whether onshore or offshore (FLNG), during the whole of 2016, and it is expected that this trend will continue until the LNG market rebalances and liquefaction costs decline.

While low natural gas prices and excess in supply of LNG will inhibit the sanctioning of LNG export projects, such as FLNGs. However, it will encourage the development of LNG import projects, such as FSRUs as these can be deployed more rapidly than shore-based terminals and hence more readily take advantage of this market situation.

3. KEY MARINE DESIGN DRIVERS

3.1 FIELD DEVELOPMENT STRATEGY

Once the field development strategy has identified a FLNG vessel as a possible exploitation option, certain fundamental aspects of the project will influence the selection of an appropriate FLNG Hull design. The field operational life, will obviously dictate the assignment of corrosion margins and the fatigue life. A harsh wave environment can induce loadings on a permanently moored structure that necessitates a new-build specially strengthened hull with extensive amounts of longitudinal material over a trading ship designed for standard Worldwide Service.

FLNG developments, like all capital intensive projects, look for as short as possible time to production in order to improve the Internal Rate of Return of the project. For oil FPSOs, the typical approach if this timescale had to be minimised and the field was in not too harsh a location, and the required life was not say above 10 years, was to take existing tonnage up from trade and convert it. However, existing LNGC tonnage is unlikely to be suitable for conversion to a FLNG, without incorporation of significant hull sponsons, due to the large deck area required for the Topsides process and liquefaction trains and associated equipments, not to mention their weight. In addition, the Capital Expenditure (CapEx) associated with the Topsides necessitates long operational lives. Hence, only in rare cases is a conversion of a LNGC appropriate, however as has been illustrated previously due to the recent downturn 3 conversions are underway and others are planned.

3.2 FIELD DEVELOPMENT STRATEGY

All of the upper deck area within the cargo region over the LNG, and potentially LPG and condensate, and adjacent water ballast tanks can be made available for mounting the Topsides namely process, power generation and utility modules. It is good practice if less hazardous Topsides plant and auxiliaries, such as utilities, power generation plant and power transformer and switchboard rooms etc., are sited between the more hazardous process modules and the accommodation block.

Due to maritime regulations, the accommodation block cannot be sited over spaces contiguous with cargo tanks. Hence, on FLNGs, and FPSOs, accommodation blocks are invariably sited over auxiliary machinery spaces containing oil burning machinery, oil fuel bunker tanks or piped in fuel gas. Obviously the size, layout, orientation and location of the accommodation block will influence the design significantly.

If a turret mooring system is chosen then this can be sited forward of the process plant if a naturally weathervaning swivel solution is adopted. If a drag chain turret system is employed then its location typically close to midships will split up the topsides arrangement. If a spread mooring system is applicable then obviously this is incorporated on the forward and aft decks, and if moored to a jetty then a standard marine mooring arrangement can be adopted which will impinge little on the upper deck.

Offshore, the offloading solution for LNG and LPG is, currently, side-by-side due to the application of 'hard-arms'. However, this is not the case for condensate where a tandem arrangement with the offtake shuttle tanker in line astern of the FLNG is typically adopted, hence the offloading system needs deck space at the aft end of the vessel for turret moored designs – for spread moored designs it is possible that a buoy system may be used.

The relative location of accommodation, process, flare, mooring and offloading systems must be considered selecting the topsides layout both from a regulatory and first principal safety case approach.

The overall Topsides layout including main power generation and process cooling water philosophies, separation (safety) area and process deck elevation requirements, craneage, maintenance ways, workshops, laydown / storage areas, access and escape routes will determine the required 'real-estate' and as such they directly drive the principal dimensions of the FLNG. Other elements that directly drive the principal dimensions etc. are discussed below.

3.3 TOPSIDES PROCESS

Specific aspects of the Topsides that drive the design from the earliest stages are:

- Relative layout and separation requirements of modules;
- Footprint of individual modules;
- Weight (dry and wet), extents etc. of individual modules;
- Longitudinal and transverse centres of gravity of individual modules;
- Vertical centre of gravity of individual modules and elevation of the process deck.

3.4 TOPSIDES EQUIPMENTS IN THE HULL

Due to the relatively large footprint required for LNG Topsides modules, then it may be advantageous from the principal dimensions, vessel balance etc. to site non-hazardous process utilities, such as sea water cooling pumps and heat exchangers, power generations /

transformers / switchboard rooms etc., below the upper deck within the hull rather than in modules on the process deck.

3.5 CONTAINMENT SYSTEM

The selection of the Containment System (CS) fundamentally drives the principal dimensions and configuration, and also structural stiffening and scantlings. The common choices of CS are:

- Self-supporting spherical Type B (MOSS);
- Self-supporting prismatic Type B (SPB);
- Membrane – typically Mark III or NO96;
- Other mature and emergent systems etc.

Each of the above have their advantages and disadvantages but these are out with the scope of this paper.

3.6 CARGO CONFIGURATION

Depending upon the required hydrocarbon products to be stored, namely LNG / LPG / Condensate, together with the selected CS and the tank sizes etc., then the configuration of the cargo region can be significantly effected as it may force all cargo tanks to be 'in-line' or 'wrapped' etc. (see Section 5).

3.7 STORAGE CAPACITY

A simple model of the production rate together with offloading frequency / parcel size(s) and required buffer storage for weather, and to a lesser degree tank inspection and maintenance requirements, will determine the minimum required LNG / LPG / condensate storage capacity. In addition the regulatory regime, chemical requirements, refrigerant (for liquefaction process) requirements etc. must also be determined and accommodated in the Hull.

3.8 ENVIRONMENT

The environment has many facets, such as: whether the location is offshore or near-shore; waves categorised as benign or harsh with unidirectional or bi-directional seas; the wind; the currents etc.

While FPSOs were first brought into operation in the relatively benign environments of the Mediterranean, Brazil and Indonesia as externally turret moored vessels they soon spread to areas with harsher environments like the North Sea. However, now there are no 'out of bounds' areas of the world's oceans for FPSOs, and hence FLNGs, just as there are no out of bounds areas for trading tankers. Even sea areas with ice etc. can be exploited using FLNGs provided suitable hull forms, construction materials and scantlings are used.

The environment imposes an upper and a lower limit on FPSO dimensions. Too small and its motions will be too extreme to provide a stable platform for the operation of Topsides. Too large and the mooring system costs will escalate dramatically.

The wave environment can also introduce large bending moments and shear forces into the hull girder. In the more extreme and harsh environments, actual longitudinal wave bending moments can significantly exceed the rule values from the classification societies for Worldwide Service.

In moderate to severe wave climates FLNGs must weathervane into the predominant wave direction to control roll motion with respect to Topsides operation and reduce mooring loads, hence a turret must be incorporated. An environment to which an FPSO would be least suited is where the vessel cannot weathervane to avoid beam seas. Such an environment could involve bi-directional seas of high energy or strong cross currents. In more benign environments a FLNG can be spread moored so that it is aligned into the predominant swell in the knowledge that non-aligned environments cannot significantly increase mooring loads or impact on motion induced downtime.

Other aspects which are typically evaluated and designed for include acceptability of motions regarding both the process plant and personnel), extreme wave loads etc.

3.9 MOORING REQUIREMENTS

Selection of the most appropriate mooring system is dictated by the environment, water depth and riser solution being employed – in fact selection of the mooring and riser solutions are directly linked, and the number and type of risers can drive the selection of the mooring system. In harsh environments the mooring loads on a FPSO can only be withstood if the vessel weather vanes into the predominant wave direction.

There is no particular constraint on deep water mooring with a monohull FLNG as the hang-off weight or tension from the mooring system can easily be compensated for with a small increase in displacement. However, a water depth of less than four times the draught will impose significant difficulties to the mooring / riser system designer due to the inherent stiffness of the system attracting high loads.

Typically, for turret mooring, the further forward the better but for a FLNG of typical oil tanker proportions will naturally weathervane if the mooring point is no more than about 10% to 12% of the Length between Perpendiculars (L_{PP}) aft of the Forward Perpendicular (FP). However, it is possible, dependent on bow form etc., that a maximum of 20% L_{PP} can be sometimes be accommodated without the need for active alignment correction via stern thruster(s). A limited degree of directional instability can be an advantage if it is necessary to use thrusters, aft and possibly forward, to maintain the vessels heading when the wind driven waves and far field swell waves do not align. This is achieved when the mooring point is 20% to 25% aft of the bow. Turret mooring systems and associated swivels, possibly thrusters and power generation, and hull reinforcement are expensive.

However, if the mooring system does not allow unlimited freedom of rotation, as with a drag chain turret, then the mooring point needs to be closer to amidships and active control, with thrusters forward and aft, incorporated in order to maintain the heading into the waves. Note that moving the turret mooring position closer to amidships reduces top end riser loads which might be a consideration in harsh environments.

Spread mooring systems are lower cost solution but can only be used safely when there is a predominant wave direction and the wave, wind and current environment that could be beam on to the vessel is relatively benign, such as off West Africa and Brazil. More advanced forms of spread mooring allow for a degree of adjustment to line tensions to modify the alignment of the vessel should the wave direction shift, but this only allows change of heading by +/- 30 degrees.

If notice of extreme environmental events such as hurricanes, typhoons or floating ice can be advised in time a Client may request a disconnectable mooring system and propulsion to facilitate the FLNG to steam away from the event, another solution may be to tow the FLNG but dependent on size the attendant ships may not be suitable.

As discussed above, thrusters can be an integral part of the mooring solution or be used only to improve the operability of the FLNG by for example maintaining best heading for low motions or providing heading control during offloading operations. If they are an integral part of the mooring solution then the Classification Society rules require high levels of redundancy in all of the power generation, control and thrusters components. Thruster installations have to consider access for maintenance at sea, as the FLNG will not be coming off station during its design life.

3.10 REGULATORY FRAMEWORK

Most of the major IACS classification societies now have rules governing the design of FLNGs. Therefore, to have a Classification Society oversee and 'approve' the hull design forms a major step in the design verification process.

In nearly all regions of the world some form of process to permit the operation of a FLNG exists, which involves audit of the design solution as well as its means of operation. This is heavily regulated by some national governments while others simply accept a Classification Society certificate as proof of safety of the solution. The regulations of some countries are prescriptive in nature whereas others have 'goal-setting' regulations that lay down requirements and require an operator to demonstrate that the risks to the operators, third parties and the environment are As Low As Reasonable Practicable (ALARP).

While the Hull designer may refer back to Classification Society rules to demonstrate the safety of their solution this has to be documented in Design and Operations Safety Cases. Specifics relating to the field environment, mooring solution and topsides and subsea interfaces have to be addressed in the Hull design Safety Case documentation.

Regulations, which are used as guidelines in the design of FLNGs can include:

- International Maritime Organization (IMO) – Safety of Life at Sea (SOLAS), International Code for the Construction and Equipment of Ships Carrying Liquefied Gases in Bulk (IGC Code), International Convention for the Prevention of Pollution from Ships (MARPOL), International Convention on Load Lines (ICLL), International Regulations for Prevention of Collisions at Sea (COLREGs), etc.
- National Regulations and Guidance Standards
- Classification Society(s) - Ship Rules, Offshore / FLNG Rules etc.
- Company Standards etc.
- Helideck - United Kingdom Civil Aviation Authority CAP437, Norwegian NORSOK C-004 etc.
- Offloading – Oil Companies International Marine Forum (OCIMF) Offshore Loading Safety Guidelines etc.
- Accommodation - International Labour Organisation (ILO) Convention, National Standards, Company Standards etc.

4. NAVAL ARCHITECTURE CONSIDERATIONS

4.1 HULL CONFIGURATION

As discussed further in Section 4.2, the configuration of a FLNG hull is driven by the environment in which it is moored, the positioning of the accommodation block, the required mooring system and risers, whether the turret / mooring is disconnectable and if the FLNG is to be self-propelled, the offloading system(s) selected etc.

4.2 PRINCIPAL PARTICULARS

As with standard FPSOs, and ships in general, the selection of the principal particulars is an iterative process and the designer needs to be mindful of the limiting relationships between length between perpendiculars (L_{PP}), breadth (B), depth (D) and scantling draught (T_S) with regard to global strength. One of the significant limiting relationships is L_{PP}/D, which relates to longitudinal strength. Too low an L_{PP}/D may result in redundant steel material, whilst too high L_{PP}/D may result in excessively large scantlings. As a guide an L_{PP}/D of between 11.0 and 12.5 is considered to be acceptable. However, unlike trading tankers or LNGCs, a FLNG is stationary and as such the L_{PP}/B ratio can be reduced to facilitate L_{PP}/D to be adjusted if necessary.

Consideration must be made as to the type of mooring system being employed as this may influence the limiting magnitude of L_{PP}/B. For example, should the FPSO be spread moored then it is feasible to have a very low L_{PP}/B but if the FLNG is turret moored then the weathervaning characteristics will be unduly affected by a low L_{PP}/B.

In simple terms, the FLNG hull must be of sufficient size and configuration for:

- Cargo (LNG / LPG / condensate) capacity and CS;
- Water ballast capacity;
- Internal turret (if applicable);
- Auxiliary machinery (Essential generators, Heating, Ventilation and Air Conditioning (HVAC), hotel services, etc.

The natural gas condensate, which is probable to be produced even from a lean gas field, is governed by MARPOL and hence it should be noted that the maximum capacity of any one condensate cargo tank is restricted to 30,000 metres[3], this is in addition to the requirement for a maximum permissible cargo tank length. The minimum number of water-tight transverse bulkheads in accordance with Classification Society rules should also be adhered to. These factors influence the maximum length of the individual cargo tanks. At the concept design phase the feasibility of individual LNG, LPG and condensate cargo tank dimensions from a sloshing aspect should be addressed.

In the case of water ballast there should be sufficient capacity to ensure a minimum draught at amidships of 2.0 +0.02L (metres) in accordance with MARPOL requirements. A minimum draught greater than this may be necessary to avoid bottom slamming in extreme environments. It is important that the inspection and maintenance of the FLNG is taken in to account at the concept design phase in order to ensure that the design is optimized in terms of layout and tankage etc.,

As a guide the positions of the peak bulkheads may be in accordance with Classification Society rules.
In general, based on the above considerations, the FLNG hull should be of sufficient length and breadth i.e. footprint to accommodate the following Topsides and Hull facilities:

- Stern offloading equipment (if applicable);
- Accommodation block size (with integrated Helideck) and location;
- Topsides Modules (process plant) footprint, weight, VCG etc.;
- Turret and turntable deck (if applicable);
- Mooring system and mooring loads;
- Well / Injection risers;
- Cooling water risers (if applicable);
- Flare Tower;
- Open deck space for mooring equipment, etc.

Of course the key to establishing the correct balance is close control of the interface between the Topsides designers, the Turret designers and the Naval Architects (Hull designers). Having established a start point it will be necessary to ensure that there is adequate space for the Topsides process plant and the turret/ risers and that the available space is usable. This of course has to be balanced against trim requirements, stability criteria and longitudinal strength requirements.

In general for FPSOs in harsh environments, it is prudent to have a higher than normal D/T ratio, hence freeboard, in order to reduce the effects of 'green' water on the upper deck. In relatively harsh environments green seas can become an issue in extreme storm conditions if this is not the case and require alternative solutions to be incorporated. Due to the relatively low density of LNG and LPG and to some degree condensate, compared to oil, it is not such an issue with FLNG designs achieving a large D/T ratio. As an illustration that ship design is an often conflicting multiple criteria problem, typically it is not desirable to have FLNGs which are too 'stiff', i.e. with a too large a Transverse Metacentric Height (GM_T), because this may result in unacceptable motion characteristics, specifically roll acceleration. This situation would be detrimental with regard to crew and process plant operability, and overall ship safety. Conversely, too low a GM_T is not just detrimental regarding initial stability but can obviously also be regarding roll angle in a seaway – it should be noted that whilst passive (e.g. bilge keels) and active (e.g. free-surface or forced roll damping tanks) solutions are available to mitigate roll, that the selection of robust principal particulars, and to a lesser degree distribution of weights (hence subsequent mass inertia as well as GM_T), in order to 'de-tune' a design from the predominant wave environment is a significantly more elegant and operationally efficient solution.

In essence the hull form of a new-build FLNG typically takes on that of a 'large box' and because there is no requirement to drive the hull through the water, very full hull forms with large block coefficients (C_B) can be used. However, the Naval Architect needs to take into account the mooring aspects and for turret moored FPSOs, weather vaning characteristics. Hence, unless the design is in very benign conditions with a spread moored system then it is typical to incorporate some shaping of the fore end, not only to reduce mooring forces but also to mitigate the effects of slamming. This shaping of the fore end can be quite crude for environments where the wind driven seas align with the swell driven seas. If there is a likelihood of a misalignment of the wind driven seas and the swell driven seas then a more refined shaping of the fore may be required in order to reduce the increased loading. The aft end generally requires some shaping, in the form of a cut-up, in order to lose buoyancy and thereby prevent excessive trim by the head. This shaping of the aft end can conveniently provide a skeg for stern thrusters (should they be necessary) and provide some directional stability for turret moored FLNGs. Should the stern not be shaped then, in general, large amounts of water ballast are required to limit the trim, which can result in excessive hogging still water bending moments and hence lead to increased scantlings. In determining the hull form the Naval Architect should aim to facilitate economic building techniques through the use of hard chines and knuckles - the use of double-curvature plate should be avoided.

4.3 MOTIONS

The motions and seakeeping performance of any moored facility, and especially a FLNG where the Topsides process plant could be to some degree motions sensitive, is of paramount importance. Therefore, the availability of accurate and comprehensive metocean date at the outset of the design is imperative in order that the Naval Architect has a sound understanding of the operational and survival environments and therefore can design the vessel accordingly.

Typical analyses undertaken include both evaluations, using 2D and 3D codes for both single and multi-body scenarios, and simulations, using full mission ship simulators. These would be undertaken not only with respect to on-site conditions but also for a pre-Topsides integration Hull only tow, FLNG tow i.e.to site and dry-dock etc. Typically, these would be to establish operational / survival / extreme motions, downtime analyses etc. in order to maximise personnel and process operability.

4.4 STABILITY

The vessel's Trim and Stability (T&S) Book also includes the acceptance criteria for maximum trim, intact and damage stability characteristics and longitudinal strength. It is normal to assess the intact stability in accordance with the IMO Intact Stability (IS) Code 2008, and the damage stability to either IMO MARPOL or IMO International Gas Carrier (ICG) Code, or in some cases the IMO Mobile Offshore Drilling Unit (MODU) Code - although the authors advise that the application of the MODU Code for FLNGs is done so with caution as the standard is significantly lower that MARPOL or the IGC Code.

All FLNGs have a Loading Instrument (LI) on-board which calculates these characteristics for various loading conditions and as such permits the Offshore Installation Manager (OIM – in maritime terms the Master) to deviate from the loading conditions given in the T&S Book. However, the most important elements of the T&S Book are the Combined Critical KG_f (CCKG) Curves [1]. The CCKG Curves are produced from the results of intact and damage stability analyses (accounting for internal configuration and various deadweight tank fills) and provide the OIM with a ready reference with regard to the loading of the FLNG. It must be noted however that the CCKG Curves do not address longitudinal strength aspects of operation, and longitudinal strength in a damaged condition is not assessed as part of the design process. The T&S and Damage Stability Books together with a Capacity Plan etc. are required to be submitted to the Classification Society for approval, in addition the LI must be approved.

4.5 SURVIVABILITY ETC.

The design of a robust vessel is paramount and hence this is assured by establishing, for the design, measures such as: extreme wave height data; immersions, bow wave (green seas) impact; emergences; slamming etc.; wind loading; snow and ice accumulations; cargo tank sloshing; collision damage etc.

4.6 LAYOUT AND OPERATIONS

As discussed above, the establishment of optimal layouts and consideration of operations is imperative. With regard to this, following are some aspects which must be considered:

- Accommodation requirements - hook up, operation, campaign maintenance;
- Helideck - number, location, fuel, parking etc.;
- Escape routes and Temporary Refuges,
- Life Saving Appliances (LSAs) - type, location, number, uplift;
- Stores transfer operations, bunkering stations;
- Mechanical handling studies;
- Laydowns etc.;
- Crew transfer operations - helideck(s), boarding;
- Control Room location(s);
- Cargo Offloading – LNG / LPG, Condensate; etc.

5. STRUCTURAL CONSIDERATIONS

The three main structural design issues for a FLNG are as follows:

- The design of the main hull structure;
- The design of the Topsides process plant support structure, considering structural continuity with the main hull structure;
- The integration of the turret.

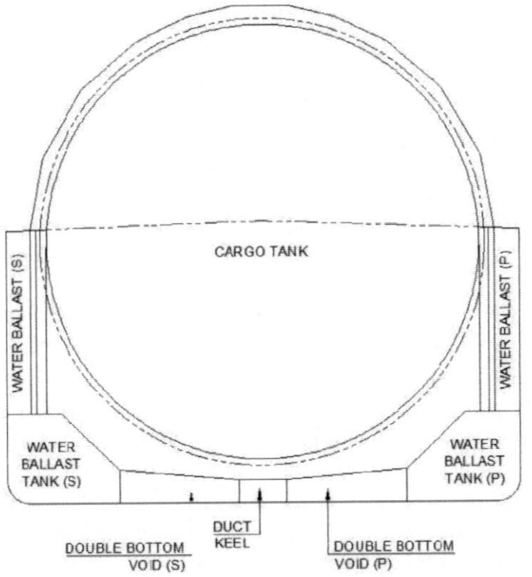

Figure 1: Self-supporting spherical Type B Configuration

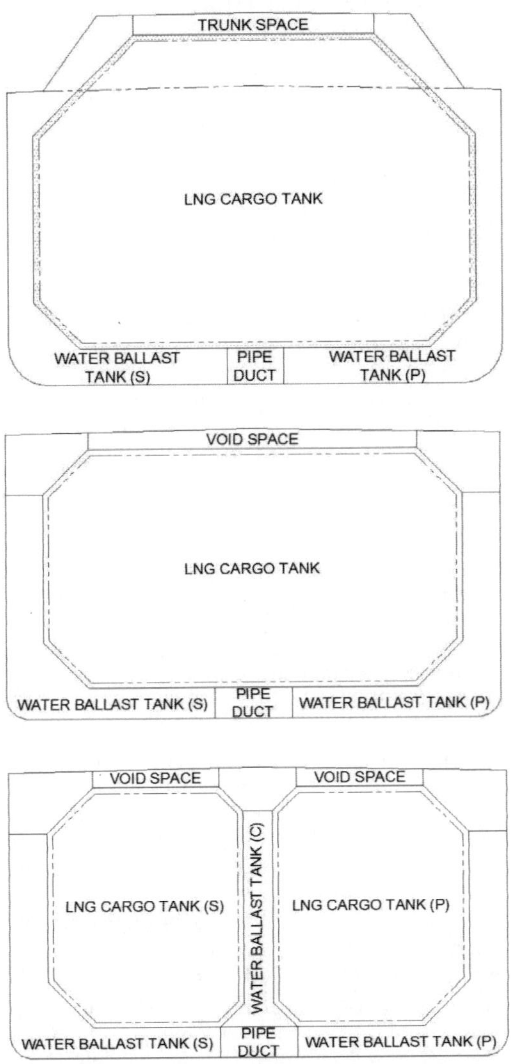

Figure 2: Three possible Membrane Configurations.

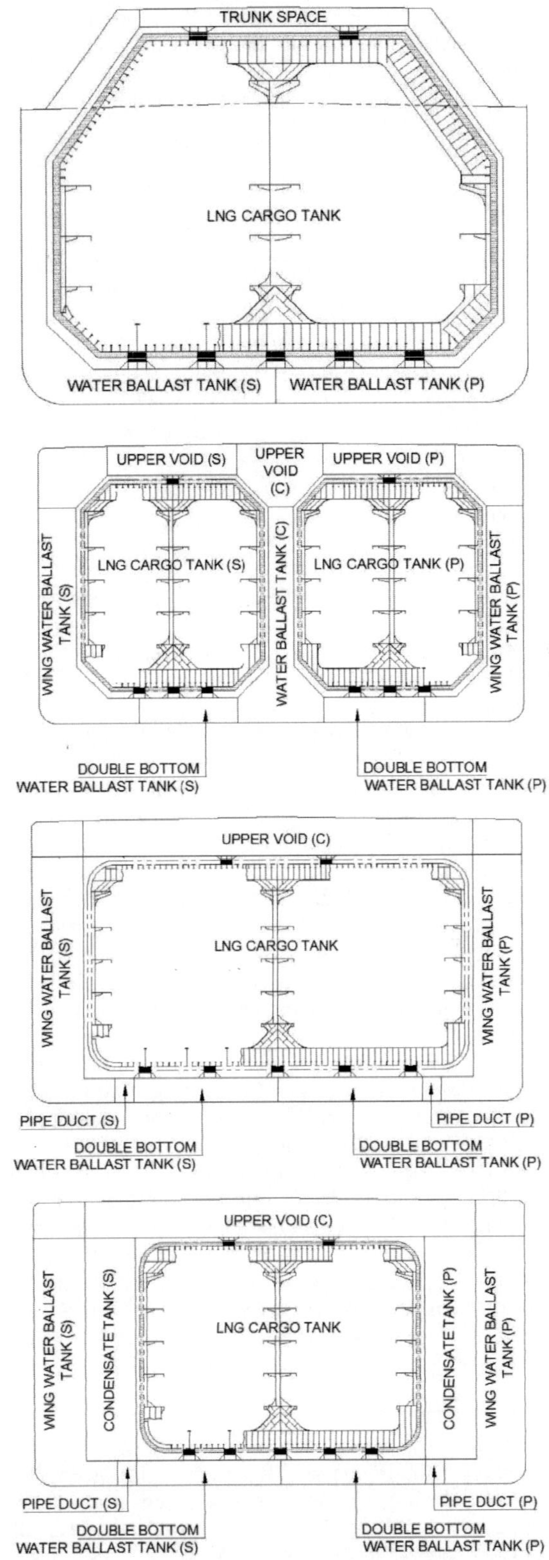

Figure 3: Four possible Self-supporting prismatic Type B Configurations

Three considerations that differ from the normal design procedure when using class rules are the need to support the Topsides process plant, the need to quantify the dynamic loads i.e. wave shear forces and wave bending moments arising from the site-specific environment, and quantification of the riser loads.

5.1 CONFIGURATION

The selection of the cryogenic CS will directly drive the global configuration of FLNG and also the structural stiffening and scantlings. Following are some examples of configurations for the three most common CSs, namely:

- Self-supporting spherical Type B (MOSS) – see Figure 1;
- Membrane – typically Mark III or NO96 – see Figure 2;
- Self-supporting prismatic Type B (SPB) – see Figure 3.

5.2 LONGITUDINAL STRENGTH, DEFLECTIONS ETC.

Once the preliminary dimensions have been selected, for the initial phases of the design, any major IACS Classification Society rules and regulations can be used to derive a set of scantlings for the midship section.

Ideally a whole ship Finite Element (FE) analysis should be undertaken using a relatively coarse mesh and include modelling of the Topsides process plant supports. From the motions analysis, accurate wave loadings can be applied to the main hull together with loads from the Topsides process plant. A smaller FE model (say 1.5 tank lengths) with a more refined mesh can then be constructed using the output data from the whole ship FEA as input loading and constraints. This smaller FE model will provide a more detailed stress / strain output and highlight regions of stress concentration such that detailed design can be undertaken. As a result, the location of the smaller model within the main hull needs careful consideration and a judgement as to the location of the most onerous region of the structure is required from the results of the whole ship FEA. Alternatively a FE model consisting of 2.5 tank lengths with a relatively fine mesh would generally meet the requirements of the Classification Society. The main issue with this is in the simulation of the boundary conditions such that the correct wave loading and Topsides dynamic loads are applied.

5.3 FATIGUE

The likely failures due to fatigue are typically due to structural detailing, for example bracket toes and in way of notches and collars. However, the structural designer should not ignore major butt welds in the upper deck plating, bottom shell and in saddle tank and hopper tank side plating.

5.4 TOPSIDES INTERFACE

While the layout of the Topsides process modules is under the control of the Topsides designers there are very important interfaces relating to the systems and support structure. The design of the systems is briefly dealt with in later Sections, therefore only some system interfaces are discussed below and those involving the support structure in Section 5.5.

Electrical interfaces are defined in the context of FLNGs as 'junction points' or 'nodes' between the Topsides and Hull installations. Depending on whether the project is a conversion or a new build and whether or not the construction is carried out at one establishment or a number of establishments, determines the number of interfaces to be considered. Examples of interfaces include:

- Hull / Topsides power generation system including the power management system;
- Hull / Topsides gas and fire detection and firefighting systems;
- Hull/Topsides to the vessel management system, including the Integrated Control and Safety System (ICSS);
- Turret ESD system;
- Cargo tank filling system;
- Topsides sea water cooling system;
- Utility services shared between hull and topsides e.g. fuel, fresh water, compressed air, instrument air, HVAC, cargo heating, inert gas, hazardous drains, heli-fuel, etc.

Even systems that have no physical connection between hull and topsides may pass through their respective space boundaries and therefore need interfaces to be managed.

5.5 SUPPORT OF TOPSIDES

While there are numerous ways of supporting Topsides modules there are generally two that are most commonly adopted, and they are:

- Module Support Frames (MSF);
- Individual Support Stools.

A MSF principally consists of external upper deck stiffened plated transverses in line with the upper deck transverse frames with vertical stiffeners at the same spacing as the upper deck longitudinals, hence giving structural continuity. The vertical stiffeners are tied into a large face plate which in turn acts as a foundation for pads or small stools to take the bolting arrangements of the module. In general there are two MSFs per Pre-Assembled Unit (PAU) or module, one at each end. The module is rigidly fixed to one MSF and is allowed to slide on the other, hence any loads arising from the hull bending are not transmitted to the module. The advantages of a MSF are:

- Even load distribution into the hull, so reduces stress concentrations and enhances fatigue life;
- Greater flexibility in dealing with late changes to module skid depths;
- Module does not experience any loads due to hull bending;
- Standard shipbuilding practice i.e. stiffened plated panels that facilitate ease of alignment.

Support stools are ideally placed on the upper deck at the cruciform of deck longitudinals and transverses but can be pyramids, and are directly connected to the modules. The individual support stools can be of cruciform construction, tubular construction or a combination of both. The advantages of Individual Support Stools are:

- Good open structure that does not allow pockets of trapped gas and gives good open access to the upper deck level;
- The potential for all stools to be made from one pattern thereby facilitating ease of production.

5.6 OTHER

Other structural considerations include:

- Topsides Crane integration;
- Flare Tower integration;
- Helideck integration;
- Raised bulwarks for Green Sea protection;
- Blast loads in emergency conditions;
- Ice Loads, etc.

6. MARINE SYSTEMS CONSIDERATIONS

The design of the hull marine piping and pumping systems shall comply with Classification Society rules, IMO SOLAS, IMO IGC Code and MARPOL, OCIMF etc.

Classification rules deal with bilge, ballast, fire and wash-deck, air and sounding pipes, fuel oil, cargo and inert gas systems giving, for example, number and minimum size of pumps, size of pipework, material types and location of equipment.

SOLAS is the generally accepted standard used to design the firefighting systems for the protection of cargo tanks, helidecks, accommodation and machinery spaces giving pump capacities, nozzle sizes, minimum operational duration, type of protection media (e.g. water, CO_2) and equipment location.

ICG Code / MARPOL deals with flooding, overboard discharges, cargo systems, etc.

In addition to system philosophies, for each of the systems diagrammatic and associated documentation, the detail of which is dependent upon the level of design i.e. either Process Flow Diagrams (PFDs) or Process and Instrumentation Diagrams (P&IDs), are required indicating the size of pipework, size of pumps, number of valves, material specification, controls and instrumentation etc. for approval by the Classification Society.

6.1 CARGO SYSTEMS

This group typically consists of the LNG, LPG and condensate systems, as required.

6.2 WATER BALLAST AND SEA WATER SYSTEMS

This group typically consists of the bilge and ballast together with the hypochlorination system, sea water fire water system, Topsides sea water cooling system etc. The water ballast system may have to be able to provide real time compensation in order to maintain constant draught and trim of the FLNG during all cargo offloading evolutions and maintain the loading of the vessel within longitudinal strength limits. In extremely harsh environments, the ballast system may be used to load the vessel in such a way as to mitigate motions etc. in storm conditions.

There are a number of broad approaches that can be considered for the design of the water ballast and hence other systems:

- 'Traditional' practice utilising a machinery space pump room housing the water ballast pumps. This system has the water ballast tanks connected to the ballast pumps with all of the associated water ballast ring main pipework interconnecting the tanks – leading to significant pipework. There would be a separate sea water cooling system for Topsides with the associated pumps located in a machinery space. In addition there would be fire pumps in the machinery space supplying to the Topsides ring main for distribution around the vessel. It should be noted that the ballast, sea water cooling and fire pumps have connections from the sea through the hull – leading to significant hull penetrations.
- 'Current' practice utilising submerged ballast pumps the number of which may not increase from the 'traditional' practice. This leads to less difficult to maintain pipework within the tanks as more pipework on sited on the open deck. All of these pumps utilise a single marine hydraulic system, which can easily provide the power required for the usually large discharge rates. The current practice on Topsides sea water cooling is to typically have submerged lift pumps within caissons, possibly located in the water ballast tanks or dedicated voids, which then discharge either via hull located coolers or direct to Topsides with the potential to save on pipework. However, there can be significant hull penetrations leading to potential problems with corrosion etc. It should be noted that all of the submerged pump options could lead to additional maintenance costs because of increased complexity.

The largest consumer of fire water systems are Topsides, therefore, current practice is to locate the fire pumps as close to the Process deck as possible. This has seen the introduction of large capacity diesel driven fire pumps located on the upper deck and fed by hydraulically operated lift pumps. This reduces the amount of pipework, the containerised system is easily installed and there are no complex exhaust routes from a machinery space.

- 'Future' practice may utilise electrically driven submerged pumps in every water ballast tank, the number of which will depend upon the required discharge rate and redundancy philosophy. These are smaller capacity pumps and as such increase the number of pumps required to maintain the discharge rates. The advantage is that there are no marine hydraulic systems and no penetrations through the hull. However, submerged electric water ballast pumps can only discharge, hence another means of filling the water ballast tanks is required – such as from adequately sized Topsides sea water cooling systems or dedicated ballast water lift pumps in caissons attached to the side of the vessel. The disadvantage is that the system costs are greater – this has to be carefully considered against any advantages gained in operations and maintenance.

6.3 FUEL OIL AND GAS SYSTEMS

These are the marine diesel oil and, if applicable, fuel gas systems.

6.4 OTHER SYSTEMS

Topsides systems can include: service water; slops; produced water; MEG; production chemicals etc.

Typically systems such as the firefighting and possibly CO_2 systems may straddle both the Hull (marine) and Topsides domains. The sea water fire water system is independent of the rest of the FLNG systems and cannot trip on barge shutdown, i.e. the diesel engines can run independently until their fuel is exhausted.

Depending on the location of the FLNG, the Heating, Ventilation and Air Conditioning (HVAC) systems for the accommodation and machinery spaces can vary dramatically and have a substantial influence on associated systems, especially the electrical power system.

Marine systems within the Hull domain are many and varied and, depending on the design premise, can include systems such as: air and sounds; inert gas; LNG void dehumidification; lubricating oil; hydraulic oil; steam; steam condensed water; demineralised water; instrument air; working air; starting air; fresh water, potable water; chilled water; sea / fresh water cooling, sewage, heli-fuel, exhaust etc.

6.5 POWER GENERATION PRIME MOVERS

Electrical power generation can be via either gas turbine, steam turbines or dual-fuel diesel prime depending on a number of parameters and preferences.

6.6 LEVELS OF REDUNDANCY

It is imperative that a consistent and robust philosophy regarding redundancy for power generation / distribution, cargo systems, ballast systems, controls is established at an early stage of design.

6.7 PROCESS COOLING WATER

As discussed above in Section 6.2, sea water cooling system for Topsides can be fed from dedicated seaboxes (mainly applicable for shallow water applications), internal caissons, hull side caissons, or a reservoir type solution.

6.8 INERT GAS PRODUCTION AND PURGING

The inert gas medium type and equipment location of this must be considered.

6.9 EMERGENCY AFFF

Considerations include areas of coverage and hence the number and location of monitors, drenchers etc.

6.10 OFFLOADING

Cargo offloading of LNG / LPG, and commonly condensate shall be via submerged or deepwell pumps. Other aspects that require careful consideration include ullage measurement, metering etc., the choice of offloading technology(s) such as hard arms, reels, floating hoses, buoys etc.

7. ELECTRICAL SYSTEMS CONSIDERATIONS

The electrical systems considerations broadly relate to the design, specification, installation and commissioning of the following systems on board the FLNG:

- Power Generation and Distribution Systems;
- Control and Instrumentation Systems;
- Internal and External Communication Systems;
- Navigation Aids Systems;
- Safety Systems, including monitoring of the hull, mooring systems, offloading systems along with the environmental conditions.

The overriding considerations in the design of the above systems are safety, availability and continuity of the systems under all operational scenarios ranging from normal production, through standby, to emergency shutdown under closed conditions and finally evacuation. While the provision of power supplies might be contractually split between the Topsides and Hull contractors, the development of the power supply and distribution system must be considered as an integrated solution and all design approvals obtained on such basis.

A number of major decisions have to be made during the determination of the required installed generating capacity and the design of the architecture of electrical supply and distribution systems. The decisions are supported by carrying out a number of principal design activities such as: the electrical load analysis; Single Line Diagram(s) (SLD), power system analysis; and system philosophies. Block diagrams of system architectures and arrangement drawings showing the locations of the principal items of major equipment and the main cable routes; etc. support the development of the hull arrangement.

7.1 LOAD ANALYSIS

The Electrical Load Analysis must take account of various operational, offloading, emergency etc. scenarios. Loads could include those from Topsides process plant, thrusters (if applicable), cooling water supply to Topsides, cargo pumps, cooling / HVAC / trace heating, marine systems, hotel loads, main fire pumps etc.

7.2 POWER GENERATION AND DISTRIBUTION SYSTEM

There is no universal method of power generation or preferred arrangement of the system architecture. The generators can supply consumers directly from one or a number of main switchboards. Transformers, downstream switchboards and distribution boards are then employed to distribute supplies to all consumers. The key SLD illustrates the overall architecture of the Power Generation and Distribution Systems. A number of generating sets can be employed to provide the power to all facets of the FLNG, a connection between the Topsides and Hull distribution systems will be present to facilitate start-up, security of supply etc. Typical power generation sources are:

- Topside Power - This is typically generated at 6.6 kV or 11 kV three phase 60 Hertz from multiple generators usually located on the process deck but sometimes, due to the large footprint of the liquefaction topsides. Prime movers are typically gas turbines but both dual-fuel diesels and steam turbines have been used;
- Essential / Standby / Black Start generators – Usually diesel generators located within the Hull, but in exceptional cases may be on Topsides, to provide power to hull systems;
- Emergency Power Generation - Usually one, but possibly two separated, diesel generators on or above the freeboard deck at a location(s) remote from the process Topsides;
- Uninterruptable Power Supplies (UPS) – to enable vital systems to remain operational when main power sources are unavailable.

8. SAFETY CONSIDERATIONS

8.1 EMERGENCY SHUTDOWN

Safety systems must be in place such as Fire and Gas Systems (FGS) and Emergency Shutdown (ESD).

8.2 STABILITY

A critical document regarding the operational safety of a FLNG is the Trim and Stability (T&S) Book, this being the primary guide to the safe operation of the vessel – see Section 4.4.

8.3 BLAST AND SPILL PROTECTION

In the design of the Topsides layout, care must be taken in determining the appropriate blast protection and incorporation of suitable safety gaps between groups of modules to prevent jet fires propagating etc. In addition the main deck etc. must be afforded suitable thermal protection in way of the modules etc. containing cryogenic liquids.

8.4 FIREFIGHTING AND PROTECTION

Firefighting systems can include:

- Sea water;
- Foam (with sea water),
- Carbon dioxide (CO_2);
- Sprinkler system / water mist;
- Portable equipment etc.

Fire protection is required to be in accordance with IMO SOLAS a Bulkhead and Deck Fire Integrity Plan is required to be submitted to the Classification Society for approval. Radiated heat from fires in the Topsides modules must be accounted for in the FLNG design.

8.5 ACCESS AND ESCAPE ROUTES ETC.

The Access and Escape arrangements are to be in accordance with IMO SOLAS. This is a minimum requirement and in general the width of corridors and stairways on offshore moored facilities is greater than SOLAS requirements (up to 1.8 metres rather than 1.1 metres). An Access and Escape Plan is required to be submitted to the Classification Society for approval.

Temporary Refuges, typically both forward and aft, must also to be incorporated in to the design of the FLNG, together with suitably sized and located Muster Stations accounting for primary, secondary and tertiary evacuation vehicles.

8.6 LIFE SAVING APPLIANCES

The Lifesaving Appliances (LSA) must be arranged in accordance with IMO SOLAS. Consideration of the commissioning, operational and maintenance manning profiles must be taken as to what is the 'baseline fit' and hence when temporary uplift of LSA is required and it's arrangement. The arrangement depends on the SOLAS class of ship, e.g. a Cargo Ship, Special Purpose Ship (SPS), etc. It is normal practice for a FLNG to be a Cargo Ship (a Class VII Cargo Ship under United Kingdom Maritime and Coastguard Agency, MCA, regulations) because the crew are generally signed-on as being fully trained in accordance with SOLAS requirements. However, if a large number of the crew are not trained then the SOLAS class would be SPS. The main implication of this is in the number and size of lifeboats required. There is also a link to the damage stability requirements because SPSs can drive the designer to Passenger Ship requirements if, for example, the crew is 240 or greater. This in turn would potentially require the application of SOLAS probabilistic damage, as opposed to MARPOL / IGC Code deterministic damage, which does not allow for cargo dropout and hence would be severely onerous for a FLNG. An LSA Plan is required to be submitted to the Classification Society for approval.

9. FUTURE EXPLOITATION AND TECHNOLOGIES FOR STRANDED

Many major offshore gas finds over the recent decades are remote from land and any market and hence transporting the gas, in whatever form, by sea can and has been proven to be a viable economic solution. The strong market for natural gas will eventually rejuvenate the development of offshore gas fields and once again the FLNG presents an attractive proposition as the three traditional technology solutions for gas recovery [2], given below, may not be viable for the reasons outlined:

- Gas Flaring - Environmentally unacceptable;
- Gas via Pipeline - Potentially expensive;
- Gas Re-injection - Expensive without revenue and can effect of any associated oil etc.

There are a number of alternative emergent technical options currently potentially available, in addition to that of LNG, for exploiting the energy potential of gas rich offshore reserves [2]. Utilising its extensive knowledge, experience and proven track record of FPSO, including Multiple Field Development Solutions (MFDS), and FLNG design, Babcock has investigated a number of these, which like LNG typically involving the transfer of mature but previously exclusively land based technologies to an offshore marine environment. Babcock has grouped feasible alternative technology contenders to FLNG in the race to process gas offshore into three broad sectors:

- Gas To Transitory Medium (GTTM) - A form of natural gas that is not stable at standard temperature and pressure (STP) such as Compressed Natural Gas (CNG, gas at high pressure typically 240 bar), Natural Gas Hydrates (NGH, ice like solids formed by the physical combination of water and gas molecules), etc. Note that LNG is also in this category;
- Gas To Liquid (GTL) – On board gas conversion to stable hydrocarbons that can, say, be blended with the crude or stored in separate tanks is an option. In this conversion, a liquid created through chemical reactions that is stable at STP, such as Synthesized Fuels e.g. Fischer-Tropsch (F-T) Diesel (C_nH_{2n+2}), Methanol MeOH (CH_3OH), Ammonia (NH_3);
- Gas To Wire (GTW) – Use of the gas to drive, typically, combined cycle generating plant, namely directly to gas turbines (GTs) and then, with steam generated from their exhaust, steam turbines coupled to electrical generators. The generated electricity is then exported ashore via subsea cables, either using AC or DC transmission depending on step-out distance, to the point of sale, typically either feeding in to the national grid or to supply directly high energy industrial customers.

Babcock have developed design solutions for a number of Floating Gas Processing, Storage and Offloading (FGPSO) vessels utilising some of the above technologies as a desk-top proof of concept. As with FLNG, a number of these designs are driven more by deck area than storage capacity as the gas processing systems are significantly larger and demand significantly more power than multi-phase well fluid separation, hydrocarbon stabilisation and gas compression systems on conventional oil FPSOs. Many of the above involve radically different FPSO solutions that are out with the scope of this paper. However, it should be noted that the Hull design can adopt a more traditional FPSO / Product Tanker practice, rather than LNGC practices followed by FLNGs.

In addition to developing traditional LPGC and LNGC designs for clients utilising all available containment systems and propulsion trains etc., Babcock has also developed a number of novel LNGC solutions with unique dimensional ratios, hull forms, configurations, gas cargos, propulsion etc. Building on this experience, Babcock has also designed environmentally robust Shuttle LNGCs incorporating Dynamic Positioning (DP) specifically to service specific FLNG developments.

In the coming decade it is more than likely that the demand for gas will rise and hence developments incorporating FLNGs will substantially increase from the current subdued levels and hence, for certain applications and fields, some of the above technologies may become accepted solutions in the field developer's 'toolbox'.

10. CONCLUSIONS

As has been reviewed in this paper, recent FLNG based concepts for offshore gas field developments have utilised various liquefaction technologies, containment systems, hull configurations, mooring solutions etc. dependent on production throughput, technology maturity, safety, distance offshore, metocean conditions, water depth etc. Vessels have ranged from giants in excess of 500 metres in length with multiple liquefaction trains giving LNG production rates approaching 10 mtpa down to solutions of 1 mtpa of around 300 metres, akin to the size of the large 2MMBbl new-build FPSOs offshore Africa and Brazil, for more marginal field developments. The flexibility of the FLNG concept is such that it can deal with these two production extremes and can be an economical field development solution for both.

With only a very few exceptions, proposed FLNGs have all utilised new-build Hulls designed for the specific process Topsides and environment. As the technology for hull design is akin to LNGC and FPSO / Oil Tanker design, it is therefore proven and most of the major IACS classification societies have Rule sets and approval procedures to cover the specifics of FLNG hull design and safety. Hence, considering the above and the fact that liquefaction technologies are mature on land, then FLNG presents a low technologically and environmentally risk solution for both operators and licensors i.e. governments.

As with the design of any marine artefact the primary goals are personnel safety, which also includes de-manning hence implicitly reducing exposure, and also asset operability. It is imperative that the Naval Architect / designer comprehensively explores the design space in an efficient manner in order to arrive at the safest and near-optimal FLNG in respect to integrity and operability. The major aspects that have to be considered in the arrival of robust principal dimensions / hull form and configuration include the process technologies (requirements and limitations), storage requirements, offloading and transportation philosophies, operational philosophies, location including metocean and climate, if the FLNG is to be permanently on-station or disconnectable (self-propelled), structural integration, redundancy requirements etc. It is imperative that the maintenance and inspection strategies etc. are inherent within the design from the concept design phase onwards. In order to ensure that a coherent and near-optimal design is developed it is crucial that the design team (process, marine, mooring, subsea, operations etc.) is fully integrated with a open communication culture in which lessons learnt are freely communicated and adopted.

The environmental issues forcing the international community and national governments to consider greener fuels could provide the impetus to develop further remote gas fields and hence start a cycle of growth and investment. Thus the FLNG, and possibly some other process technologies, may well start to take centre stage in the FPSO market in the next decade.

11. DISCLAIMER

The views expressed in this paper are those of the authors and do not necessarily represent those of the organisations with which they are affiliated and the professional institutions of which they are members.

12. REFERENCES

1. HUTCHINSON, K.W. and SCOTT, A.L., 'Tanker Damage Stability: What are the Issues and Solutions?', *International Conference on the Design and Operation of Tankers*, The Royal Institution of Naval Architects, (Eugenides Foundation) Athens, Greece, 8th and 9th June, 2011.

2. MACKIE, G.C., HUTCHINSON, K.W. and WANLESS, D., 'Exploiting Stranded Gas Reserves: Options and Solution Development', *Deep and Ultra Deep Water Offshore Technology, International Conference*, The Institute of Marine Engineers and the University of Newcastle upon Tyne in association with the Society for Underwater Technology, Newcastle upon Tyne, United Kingdom, 25th and 26th March, 1999.

13. AUTHORS BIOGRAPHIES

Keith W Hutchinson is Engineering Manager and acting Chief Engineer with Babcock Energy and Marine Technology at their Tyneside offices in North Shields, United Kingdom, where he undertakes commercial and offshore marine consultancy for a range of clients globally. He is responsible for the coordination and management of all engineering teams and, as senior Technical Authority, also for all governance aspects relating to whole Ship Design, engineering etc. He is Subject Matter Expert for the Naval Architecture domain across all offices in the United Kingdom.

Keith leads multidisciplinary teams formulating design solutions, together with cost estimating and investment appraisals. Over an almost 30 year career as a ship designer he has been constantly at the forefront of the technical development of robust but flexible concept / basic designs for a multitude of extremely novel / complex / bespoke merchant, offshore and naval vessels, both new-build and conversions - from sub 20 metre high speed catamarans / landing craft to FPSOs / FLNGs in excess of 500 metres for the harshest of environments - consistently striving with clients to explore unknown multiple dimensional design spaces.

Keith graduated in Naval Architecture and Shipbuilding from Sunderland Polytechnic in 1990 and from the University of Newcastle upon Tyne in 1992 with a Bachelor of Engineering in Marine Technology with First Class Honours in Naval Architecture.

Keith worked in a number of shipbuilding and shiprepair yards on the River Tyne, including Swan Hunter Shipbuilders, and upon joining his current employer in 1995, then Armstrong Technology, he was seconded to the Engineering Design Centre at Newcastle University for over three years as a Research Associate developing multiple criteria optimisation / decision support tools for the design of RO-RO passenger ships and other vessels.

Keith has published over 30 technical papers and articles on naval architecture and ship design etc., in addition to being invited to give many presentations. Keith sits on a variety of professional technical committees and is active in the organisation of a number of professional learned institutions. He is a Chartered Engineer, Chartered Marine Engineer, Fellow of the Royal Institution of Naval Architects, Fellow of the Institute Marine Engineering, Science and Technology and Fellow of the Society of Naval Architects and Marine Engineers.

David C Dobson is Programme Director with Babcock Energy and Marine Technology based at their Tyneside offices in North Shields, United Kingdom.

His career began in 1973 as a Ship Draughtsman with Swan Hunter Shipbuilders on the River Tyne. He attended Newcastle College, Sunderland Polytechnic and Newcastle University gaining a Bachelor of Science with Honours in Naval Architecture in 1983. He quickly joined the management team in Swan Hunter taking up various positions in production, production engineering and project management. He gained a Masters of Business Administration in Strategic Management in 1992. He left Swan Hunter in 1993 and subsequently worked in a variety of Offshore, Engineering and Shiprepair companies as General Manager and Director. He joined Babcock in 2007 and is currently Programme Director for the commercial marine and oil and gas business stream.

David is a Chartered Engineer and was awarded Fellow of The Royal Institution of Naval Architects status in 1996.

DESIGN & CLASSIFICATION OF IMABARI 178,000M3 LNG TANKERS

H Inoue and **T Nakatani**, Imabari Shipbuilding Company Limited, JAPAN
K Yuasa and **H W Chung**, Lloyd's Register Group Limited, JAPAN

SUMMARY

This paper will introduce the Imabari 178,000m3 capacity LNG tanker fitted with GTT MKIII-Flex membrane containment system with 0.1% BOR and powered by dual fuel diesel engines directly driving twin propellers optimised for fuel efficiency. The ship has been designed for trans-Pacific shipment of LNG via the new Panama Canal which opened in June 2016. Imabari's and Lloyd's Register's perspectives will be presented, highlighting the salient design features and measures to ensure stringent standards in safety, serviceability and performance are met; including HAZID/HAZOP of high pressure fuel gas supply system, strength and fatigue verification, noise analysis and vibration assessment of the hull and containment system. When the first ship of the series is delivered in 2017/Q2, it will be amongst the largest and most energy efficient LNG tankers able to transit the Panama Canal, and the largest membrane LNG tanker constructed in Japan up to that point, setting a new standard for the LNG shipping industry.

1. INTRODUCTION

Imabari Shipbuilding Company Limited have been building LNG Tankers fitted with GTT MKIII membrane containment system since 2008, and established the process of constructing safe and reliable LNG tankers.

In 2015, global LNG trade accounted for about 240 million tons, equivalent to about 30% of world gas trade and 10% of world gas consumption respectively [1]. 30 countries were importing LNG with Asian countries accounting for 75% of LNG imports over the last decade, and 19 countries were LNG exporters. On the supply side, the Middle East was the largest exporting region with 40% of global LNG exports.

On the supply side, however, the introduction of hydraulic fracturing technique has enabled natural gas to be extracted from shale rock formations across the United States. This revolutionary technology created a surge in natural gas supply which exceeded domestic demand, creating very favourable conditions for the US to be transformed into a net exporter of natural gas, and more LNG tankers will be needed for transporting LNG from the US to gas consumer countries mostly in Asia.

IMABARI responded to this emerging market with the introduction of a new LNG tanker design optimised for trans-Pacific shipment of LNG from the US through the new Panama Canal locks which have started operations in June 2016.

2. FEATURES OF IMABARI 178,000M3 LNG TANKER

The high level specification of the new Imabari 178,000m3 LNG tanker is "the most advanced LNG tanker design", with the following key objectives:

- to achieve the lowest fuel consumption,
- to enable easy and flexible operation, and
- to ensure highest reliability by adopting proven technology.

In order to realise these objectives during the design development, the designers utilised valuable experience and knowhow obtained from construction and in-service feedback of the previous 154,000m3 GTT MKIII LNG tanker series and that from building various ship types at the IMABARI group shipyards, which is currently the largest in Japan and amongst the top 3 in the world.

The ship also offers many environmentally friendly features in the construction and operation including minimal discharge of CO_2 by adopting the most fuel efficient propulsion system, eligibility for LR's "ECO" environmental class notation, including Inventory of Hazardous Materials (IHM notation) on-board for ship recycling and adopting CO_2 as foaming agent for the fabrication of cargo tank insulation.

The principal particulars are summarised in Table 2.1.

Table 2.1: Principal Particulars of Imabari 178,000m3 LNG Tanker

	Particulars
Length (overall)	abt. 296.9 m
Breadth (moulded)	48.70 m
Depth (moulded)	27.00 m
Design Draught (moulded)	11.55 m
Scantling Draught (moulded)	12.60 m
Cargo Containment System	GTT MK III Flex
Boil-Off Rate (/day)	0.10 %

MAIN ENGINE	
Type & no of set	7G70ME-C9.2-GI (2 sets)
MCR, kW, rpm	17,180 kW (70.0 rpm)
Service Speed	Abt. 19.5 knots

2.1 PRINCIPAL DIMENSIONS AND HULL FORM

Principal dimensions are decided based on the conditions of 1) new Panama Canal transit and 2) compatibility at major international LNG terminals with size limitations. So the following principal dimensions were selected:

- Length overall to be 296.9m, less than 300m
- Breadth to be 48.7m, less than 49.0m, new Panama Canal requirement
- Design draught to be 11.55m, to meet major international LNG terminals' draught restriction.

With these principal dimensions established, the design aims to maximise the cargo capacity and optimise energy efficiency by adopting twin-skeg/twin-screw configuration to obtain the best combination of hull and propulsive performances despite normally higher initial cost than single-screw design.

Twin screw design was selected not only for reasons of performance but also out of consideration for the safety aspect of propulsion system redundancy.

For large LNG tanker designed with consideration of draught restriction of terminals, twin-screw ship is more efficient compared to single screw ship. Frictional resistance increment caused by increasing wetted surface area, which is a disadvantage of twin-skeg, is unavoidable. However, by adopting optimized high performance aft hull form and improvement of propeller efficiency due to reduced load on one propeller, twin-skeg ship can overcome the disadvantage and achieve less required power compared to single screw ship.

Forward hull form, especially around the bulbous bow, was designed with minute attention to minimize wave resistance in ballast condition and in actual operational speed range as well as in loaded draught.

To achieve higher propulsion efficiency, a rudder bulb is installed on the leading edge of each rudder just after the propeller cap. It recovers energy loss due to propeller rotation. Initial rudder angle was also optimized to improve propulsion efficiency.

Owing to these features, satisfactory performance and low fuel consumption was achieved in spite of the wider breadth and shallow draught.

To maximize cargo tank capacity and volumetric efficiency, the world's first trapezoidal No.1 cargo tank was adopted in Imabari's 154,000m3 LNG tanker series. The new 178,000m3 LNG tanker is also arranged with trapezoidal No.1 cargo tank. The advantages of adopting the trapezoidal tank can be summarised below:

- No.1 cargo tank capacity increased by about 20 %,
- Improvement of structural continuity,
- Improved trim condition during laden voyage, and
- Increased flexibility to optimise the hull form.

The configuration of trapezoidal tank is shown in Figure 2.1.

The distance between cargo tanks and outer shell complies with the more stringent requirement of the revised IGC code as a voluntary design consideration to enhance the ship's safety.

2.2 CARGO CONTAINMENT SYSTEM

Latest GTT MKIII Flex membrane containment system with BOR of 0.10% per day is selected for increased robustness and to achieve greater energy savings.

The MK III system had been well established since 1993 and is a cryogenic liner directly supported by the ship's inner hull. This liner is composed of a primary metallic membrane positioned on top of a prefabricated insulation panel including a complete secondary membrane.

The primary membrane is made of 1.2mm thick corrugated stainless steel 304L. It contains the LNG cargo and is directly supported by and fixed to the insulation system. Standard size of the corrugated sheets is 3m x 1m. This membrane is folded to create bi-directional corrugations. Knots at the intersections of the corrugations allow deformation due to thermal extension/contraction and hull structural movement to be absorbed.

The secondary membrane is made of triplex, a composite laminated material made up of a thin sheet of aluminium sandwiched between two layers of glass cloth and resin. It is positioned inside the prefabricated insulation panels between the two insulation layers, and bonded at panel joints with flexible secondary barrier sheet.

While a few incidents have been reported with some of the earlier MK CCS, such as secondary barrier tightness and sloshing damage, causes of the underlying issues have been well understood and countermeasures and design improvements approved by classification societies for the updated MK III systems.

The reliability of Imabari MK III containment system has been corroborated by good service records of Imabari 154,000m3 LNG tanker series since their delivery in 2008.

There is no fundamental change from the original MK III system to the MK III Flex system, except the thickness of reinforced PUF of secondary insulation is increased to achieve low BOG of 0.1% per day as shown in the Figure 2.2.

Automatic bonding is widely used and the latest flexible bonding material is applied for secondary barrier connections to maintain the quality of its tightness integrity.

The primary membrane corrugation had been reinforced against sloshing impact pressure by introduction of ribs and wedges where necessary, as illustrated in Figure 2.2, and the design has been approved by Lloyd's Register with reference to their Guidance Document [2].

Imabari and LR with support of GTT are paying continuous close attention to the CCS at every stage from supplier certification, to installation, to testing and commissioning, thus ensuring the MK III Flex cargo containment system meets the most stringent quality standard throughout the construction of the ships. An aspect of the CCS quality scheme is the implementation of "Construction, Testing and Installation Plan" or CTI Plan required in LR's Gas Ship Rules [3].

2.3 PROPULSION, ELECTRIC AND STEAM GENERATING PLANT

This vessel is fitted with two (2) sets of low speed 2-stroke diesel cycle dual fuel engine called "ME-GI" engine with the following environmentally friendly features.

- Higher engine efficiency than Otto cycle dual fuel engine by 3 – 5%.
- Higher responsiveness to load fluctuation to minimise propulsion energy loss by coupling with large diameter fixed pitch propeller without any devices between engines and propulsion shafts.
- Methane slip is a negligible concern due to proper burn-out in engine combustion chamber, virtually eliminating discharge of unburnt methane to the atmosphere.

In addition to the main propulsion plant, one (1) set of Shaft Generator on each propulsion shaft and one (1) set of Exhaust Gas Economizer (EGE) system on each exhaust pipe of ME-GI are installed in order to meet electricity and heating requirements efficiently by running the ME-GI engines at normal sea going conditions without using other electric generation systems and auxiliary boilers.

This vessel is also fitted with four (4) sets of dual fuel generator engines and one (1) set of dual fuel auxiliary boiler to ensure flexible operations at loading/discharging, slow steaming etc. These can be seen in Figure 2.3.

2.4 FUEL GAS SUPPLY SYSTEM

This vessel is equipped with High Pressure Fuel Gas Supply System (HP-FGSS) for the main engines (ME-GI) and Low Pressure Fuel Gas Supply System (LP-FGSS) for Dual Fuel Generator Engines (DFG) and Dual Fuel Boiler. Re-liquefaction System (RL system) and Gas Combustion Unit (GCU) are also installed to provide flexible and economical BOG treatment at all operating profiles with redundancy.

The Fuel Gas Supply System and BOG Handling System is shown in Figure 2.4

The purpose of HP-FGSS is to supply fuel gas to the ME-GI engines, which require high pressure, typically 25-30 MPa depending on demand, at ambient temperature up to 45ºC. HP-FGSS consists of HP Compressor System and HP-Pump System. HP-Compressor System consists of 5-stage reciprocating compressor which can process gas from cryogenic to ambient condition, heat exchangers to control temperature, flow control devices and necessary auxiliaries. HP-Pump System consists of cryogenic reciprocating pump with variable speed function to control flow, HP-Vaporiser to vaporise and control temperature, and necessary auxiliaries.

In the RL system, the BOG is cooled and condensed to LNG in a cryogenic heat exchanger whereby cryogenic temperature is produced by means of a nitrogen compression-expansion cycle. RL system consists of 3-stage centrifugal compressor, cryogenic heat exchanger, Nitrogen loop with N2 Compander and necessary auxiliaries.

For LP-FGSS, a 3-stage centrifugal compressor (LPC) which is part of RL system can be used. Also, a side stream from 2nd stage discharge of HP-Compressor system can be used as LP-FGSS.

The typical operating profiles are as follows:

<u>Laden Voyage</u>

BOG is compressed to about 30MPa by HP-Compressor (5-stage, reciprocating) and sent to the ME-GIs. A side stream can be taken out after 2nd stage of HP-Compressor for sending to low pressure gas consumers such as DFG, GCU and DF boiler. In case there is any BOG in excess of what is required for consumption by the ME-GIs, or not utilised in the ME-GIs, the gas is liquefied in the RL System and returned to the cargo tanks.

If additional gas is required by the ME-GIs, supplemental gas is fed by the HP-pump system. When operating the RL System, BOG is fed by dedicated centrifugal BOG compressor with typical discharge pressure of 0.8MPa. BOG compressor can also supply gas to low pressure gas consumers.

<u>Ballast Voyage</u>

Available BOG is compressed to 30MPa and fed to the ME-GIs. Additional gas for ME-GIs is fed by HP-pump system or Forcing Vaporiser system. On ballast voyage, RL system is normally not in operation.

Anchoring/transiting Panama Canal

When there is minimal or no gas consumption by the ME-GIs, BOG is liquefied in RL system and returned to the cargo tanks.

3. CLASSIFICATION ASPECTS

The ship is designed to comply with Lloyd's Register (LR) Rules incorporating the IMO IGC Code (effective before 1 Jan 2016) and the USCG 46CFR Part 154 for worldwide operations other than Alaskan waters. The following notations and descriptive notes are assigned to the ship.

✠100A1 Liquefied Gas Tanker, Ship Type 2G, Methane (LNG) in Membrane Tanks, Maximum Vapour Pressure 0.25 bar and Minimum Cargo Temperature minus 163°C, ShipRight (SDA, FDA plus(40, WW), CM, ACS(B)), ✠LMC, UMS, CCS, ICC, NAV1, ✠Lloyd's RMC(LG), ECO(BWT, IHM), *IWS, LI

With the descriptive notes "ShipRight (SERS, SCM, BWMP (T, S)), ETA"

The notations SDA and FDA stand for "Structural Design Assessment" and "Fatigue Design Assessment" respectively. These indicate that the design has been assessed in accordance with the ShipRight SDA [4] and FDA [5, 6] procedures. "FDA plus(40,WW)" indicates that the fatigue lives have been verified for 40 years worldwide trade. "CM" notation, which stands for Construction Monitoring [7], is mandatory where the SDA and FDA notations are assigned. It requires the critical structural locations identified by the SDA and FDA assessments to be identified in a "CM Plan" with enhanced alignment standards approved by LR, to be adhered to and specially surveyed during construction. The CM Plan is maintained on board the ship. While the "Ship Structure Access Manual" (SSAM) now mandatory as part of the SOLAS PMA requirement for oil tankers and bulk carriers are not applicable to gas carriers, in a sense the CM Plan is a more stringent precursor which was introduced by LR since the mid-1990s.

4. STRUCTURAL DESIGN VERIFICATION

4.1 STRUCTURAL DESIGN ASSESSMENT

A full length full breadth finite element model (FEM) was used to evaluate the primary structure of the ship in order to ensure adequate yield and buckling strength against the static and dynamic loads. The inner hull stresses were also evaluated to ensure they do not exceed the containment system designer's specification. Individual critical structural locations were evaluated by fine mesh FEMs to capture the stress concentration and to ensure that the stresses are within allowable limits. Figure 4.1 illustrates the global FEM (showing port side only to reveal inner hull structure) and local fine mesh areas evaluated. Some of these areas are required by the Class procedures, and some are selected in accordance with project specific requirements.

The areas examined by detailed FEM are given below:

(1) Lower and upper chamfer hopper knuckles at mid-tank length of the cargo tank No.2:
- The inner bottom connection to hopper slant plate (lower knuckle connection of the lower chamfer)
- The inner skin connection to hopper slant plate (upper knuckle connection of the lower chamfer)
- The inner skin connection to upper chamfer slant plate (lower knuckle connection of the upper chamfer)
- The inner trunk deck connection to upper chamfer slant plate (upper knuckle connection of the upper chamfer)

(2) Connection of cofferdam bulkheads to inner bottom between No. 2 & 3 cargo tanks.

(3) Connections of cofferdam bulkheads to side structure including horizontal girder between No.2 & 3 cargo tanks:
- Transverse bulkhead stringer No.15
- Side stringer No. 12

(4) Scarphing at aft end of trunk deck into superstructure:
- End of sweep bracket;
- Selected window and door openings.

(5) Liquid dome opening:-
- Corner of openings on upper trunk deck;
- Corner of openings on inner trunk deck:
- Coaming brackets end.

(6) Scarphing at fore end of trunk deck into upper deck.

The Cargo machinery rooms integration into trunk deck structure was examined according to coarse mesh FEM stress criteria and a fatigue check was carried out using the nominal stress and appropriate joint class.

4.2 FATIGUE DESIGN ASSESSMENT

The FDA procedures were originally developed for mandatory application to oil tankers and bulk carriers in the early 1990s to address specific fatigue issues for those ship types. For LNG tankers, which do not generally manifest fatigue related issues compared to tankers and bulk carriers, the FDA notation has been offered as an option. It has in fact been a de-facto standard for LNG tanker projects since the first application in the late 1990s. The "FDA plus (years, WW or NA)" notation was subsequently introduced to meet a demand by designers and owners to give formal register book recognition to a fatigue standard above that required for the basic FDA notation.

Longitudinal stiffener end connections to primary members on the outer and inner hull are assessed by the "FDA Level 2" approach. FDA Level 2 applies a spectral fatigue assessment approach using simple beam models, stress concentration factors, parametric loads and motions data, and wave scatter data derived from voyage simulation [8].

Primary member structural connections are assessed by the FDA Level 3 approach. In order to be eligible for the FDA plus notation, the ship is required to be examined by the more rigorous FDA Level 3 procedures which utilise a first principles based spectral fatigue approach in conjunction with hot spot stress obtained from very fine mesh FE models and ship motions and wave pressure obtained from a full 3D hydrodynamics simulation, together with the wave scatter data from voyage simulation derived from FDA Level 2 assessment [8].

Generally all high stress locations assessed for fine mesh stress analysis are also selected to be assessed by FDA Level 3 for fatigue strength. The very fine mesh model used for assessing the trunk deck aft scarfing arrangement is shown in Figure 4.2 for illustration.

4.3 HULL SLOSHING ASSESSMENT [2, 9]

The inner hull scantlings are verified in accordance with LR's procedures using the in-house AQUARIUS software, which employs Openfoam algorithms. A snapshot of the fluid motion from the simulation at 70% fill level under transverse plane motions is shown in Figure 4.3.

5. NOISE AND VIBRATION ASSESSMENTS

5.1 GLOBAL HULL VIBRATION ASSESSMENT

The primary objective of the analysis is to investigate the global dynamic behaviour of the vessel, and to confirm the acceptability of the global vibration characteristics by examining global hull vibration modes and responses to propeller and engine excitation, and to evaluate the velocity response levels against the criteria permissible values [10, 11].

As part of the applied dynamic loads, propeller forces based on model tests were used in the analysis. Propeller hull surface forces were calculated using CFD based on propeller dimensional data and wake data from model tests.

Seven engine load cases are considered at 100% MCR, these being:

- Order 1 engine external and guide force moments.
- Order 2 engine guide force moments.
- Order 3 engine guide force moments.
- Order 4 engine external and guide force moments.
- Order 6 engine guide force moments plus propeller 1ST blade loads.
- Order 7 engine guide force moments.
- Order 12 engine guide force moments, plus propeller 2ND blade loads.

Dynamic loads from the Cargo Machinery Room, which lies near the front of the accommodation block, have also been included. Based on the compressor shaft speed, the 1ST Order compressor loads are included in 12TH Order of the engine/propeller loads.

Two sailing conditions are investigated:

- Loaded Condition
- Sea-trial Condition

The global hull vibration analysis makes use of the finite element method for the calculation of the structural natural frequencies and subsequent responses at requested locations. The finite element model used in the Structural Design Assessment was modified to suit the requirements for Global Hull Vibration Analysis.

Velocity responses are calculated at the following locations.

- Navigation Bridge – Centre forward wall
- Navigation Bridge Wing tip Port and Starboard
- Engine Casing Top Port and Starboard
- Funnel Top – Forward Centre
- Transom Top – Centre

The dynamic loads acting on the vessel originate from two sources.

- At the engine foundations and bracings resulting from vertical, horizontal, X-Type guide force and H-Type guide force engine moments.
- On the hull above the propellers, at the stern tube bearings and at the thrust bearings due to forces generated by propeller rotation.

The propeller load components are applied in phase, to represent both port and starboard propeller blades acting in unison, and as a separate case, anti-phase where the starboard propeller load components are reversed, except for the thrust bearing X load.

Velocity responses are of interest at two discrete engine speeds, that is, MCR and NCR respectively. The responses are calculated for a frequency range covering ±20% of the frequencies of interest, and that these should not exceed the permissible values within this range. The upper bound response output frequency limit for MCR is MCR+5%.

The assessment criteria are based on ISO6954-1984E and ISO6954-2000E. For 'normally occupied spaces', that is the bridge centre, criteria are adopted to ensure comfort of inhabitants, and for steel structures to avoid structural damage. They are based on the following levels:

Workspaces or regularly habited areas,

- 1 Hz – 5 Hz : 20 mm/s to 4 mm/s.
- 5 Hz and above: 4 mm/s.

Other areas to avoid structural damage,

- Maximum acceptable velocity 30 mm/s.

Figure 5.1 shows the incorporation of the pump tower and engines into the global FE model and first non-rigid body natural vibration mode in loaded condition.

5.2 LOCAL PANEL VIBRATION ASSESSMENT

The primary objective of the local panel vibration analysis was to identify possible resonance of local panel natural frequencies with propeller first and second blade excitation frequencies for the Accommodation Decks, Aft Peak and the Engine Room.

The natural frequencies for selected panels in the accommodation decks have been analysed in air, with outfit weights applied as uniformly distributed masses on internal areas of the decks. For the accommodation decks, the analysis makes use of the finite element method for the calculation of the structural natural frequencies. The FE model as used in the SDA analysis was modified to suit the requirements for local panel vibration analysis by modelling each deck in a mesh size with sufficient fineness to capture the panel natural modes.

For the aft peak and engine room areas, the frequencies were evaluated using analytical techniques for comparison with first and second order propeller blade frequencies. The natural frequencies for selected panels were calculated in air, and with fluid on one side or both sides as appropriate for their locations.

All accommodation deck panels are considered to be acceptable with respect to local panel vibrations. Estimated natural frequencies for selected aft peak and engine room panels also indicated all panels pass the avoidance criteria.

5.3 PUMP TOWER VIBRATION ASSESSMENT

Vibration analysis of the pump tower was carried out in two steps [12]:

- Normal mode analysis of pump tower finite element model along with suitable constraints to represent the dome top support and baseplate support.
- Incorporated pump tower model into global finite element model, see Figure 5.1. Normal mode and vibration (frequency) response analyses were carried out, and the velocity response levels were assessed against permissible criterion.

Analysis was carried out for the pump tower in the same two conditions selected for hull global vibration assessments.

In wet condition, influence of fluid external to pipe structure was taken into account in the form of added extra mass to pipes. These masses have been calculated from the standard formulation for cylinders, and added to the FE model as lump mass at the nodal points. Fluid trapped in the pump tower's three main columns is also included.

For the stand alone pump tower finite element model, it is recommended that predicted natural frequencies in longitudinal, transverse and torsion modes are to be outside the ±20% of propeller blade excitation range. When the pump tower is included in the global hull model with top and lower supporting structures modelled, it is considered that the dynamic behaviour of the pump tower model is more accurate than pump tower model analysed independent of the global hull. Based on this consideration and to be in line with hull global vibration analysis, the recommended frequency avoidance range is between 10% below NCR and 5% over MCR.

To avoid vibration induced structural damage, LR's recommendation is that single peak velocity response levels in all longitudinal, transverse and vertical directions is not to exceed 30mm/s in recommended frequency avoidance range. However, for the pump tower structure, which is considered to be one of the critical components in a LNG tanker, a lower permissible vibration criterion of 15mm/s is adopted. This criterion may be more stringent than that applied by the CCS designer and other class societies.

5.4 NOISE ASSESSMENT [11, 13]

The objective was to evaluate to what extent the proposed design and the layout of the ship would comply with the agreed noise limits in order to minimize harmful noise exposure and increase comfort of the crew. Both structure-borne and airborne noise calculations were performed by LR Consulting. The prediction is based on information such as General Arrangement, accommodation plan, built-up plans, ventilation data and machinery particulars; which are required for the assessment. These documents include the Vendor's noise and/or vibration data for some equipment.

The noise limits in the accommodation and machinery areas refer to IMO Resolution A.468(XII) as per project specifications. The rules are in terms of overall noise levels in dB(A) (noted LAeq) and are derived from calculated octave band noise spectra. Two operating conditions were considered in this project, the oil fuelled and the gas fuelled conditions.

Structure-borne noise is basically a vibro-acoustic problem. It is noise transmitting through the ship's steel structure and then radiated into each receiver room via the floor, bulkheads, windows and ceiling. Sources of structure-borne noise are major machinery such as main engines and propellers or equipment located in the vicinity of the accommodation.

Structure-borne noise levels at each receiver position were predicted using a hybrid calculation software combining a transfer function method and SEA (Statistical Energy Analysis) techniques. The software has been continuously developed over the past years and has been validated against numerous sets of measurement data, e.g. sea trial data for different types of vessels.

Each identified source of vibrations was placed in the model based on their coordinates and tied to its deck. The input of vibrational energy is represented using vibration velocity levels below the equipment feet. Velocity levels depend on the mounting type (rigid or resilient) of the equipment and foundation mobility (or stiffness). These were either provided by Vendors or derived from LR's extensive in-house database. Transmission and attenuation of vibrational energy in the steel structure was modelled in all three directions.

The receivers were incorporated based on their coordinates and dimensions. The extent of radiated structure-borne noise depends on the room built-up properties. The room acoustic properties (reverberation time) also affects the sound field and therefore the resulting noise levels. The model copes with all types of floors and panels. All construction elements are defined with an acoustic radiation function based on measurements. Special purpose floors such as viscoelastic and floating floors are introduced and modelled in specific rooms in close vicinity of machinery.

Airborne noise calculations were performed in rooms containing noisy equipment and in rooms located in vicinity of high noise areas. For this purpose, a specially adapted commercial room acoustic software was used. The software is based on a hybrid image source and ray tracing calculation method, enabling fast and reliable calculations. Validations of the results have previously been carried out against in-house noise data, e.g. for marine applications and leading to small deviations in the order of 1-2 dB.

The Vendor's sound power data were used as input in the different machinery areas. The results of the software are given in terms of noise contour map in the room, plus a handful of other room acoustic parameters. Figure 5.2 shows the noise contour in the Cargo Machinery Room. These results are compared against the noise limit in the room and were also used for sound transmission calculations to surrounding areas. Airborne transmissions via floors and bulkheads were computed using the sound reduction index function of the considered construction.

External noise at outdoor receivers such as listening posts (bridge) were calculated based on Geometrical acoustic theory and Vendor's sound power data of exhaust outlets.

The noise prediction study showed compliance with IMO's noise requirements in both operating conditions. Steel floating floor designs are applied on B-deck rooms to mitigate structure borne noise from the running HP compressor located on the trunk deck.

6. RISK ASSESSMENT FOR FUEL GAS SUPPLY SYSTEM

6.1 GENERAL

High pressure fuel gas supply system (HP FGSS) is introduced to supply fuel gas at high pressure (30 MPa) to the ME-GI engines in order to achieve the most economical fuel gas consumption, and risk assessment for this system is considered essential to establish that the level of safety is comparable to conventional LNG tankers with LP FGSS, even though most of the FGSS components are selected from major suppliers.

Moreover, Class and Statutory Rules and Regulations increasingly require risk studies to be undertaken to identify hazards and to assess and control the corresponding risks for novel designs, such as LR's Rules Pt 7, Ch 14 "Requirements for Machinery and Engineering Systems of Unconventional Design". Typically, such studies are required for designs which deviate from existing rule and regulations or for novel designs for which prescriptive rules and regulations do not currently exist [14].

6.2 PROCESS FOR RISK ASSESSMENT

To ensure that such studies are undertaken consistently, with an appropriate degree of rigour and in a manner consistent with applicable Classification and Statutory requirements, risk studies associated with the use of low flash point fuels (e.g. LNG) should be undertaken according to the four stage process shown in Figure 6.1, namely:

Stage 1: Design and Safety Statement
Stage 2: Risk Assessment (HAZID, HAZOP)
Stage 3: Revision and Supporting Studies
Stage 4: Final Design Assessment

Responsibility for conducting the studies for each of the stages lies with the organisation requesting approval or approval-in-principle (AiP) from LR. It should be appreciated that the assessment process aims to ensure that all reasonably foreseeable hazards associated with a particular design are adequately controlled, irrespective of whether they may eventually fall within the scope of Classification approval or within the scope of Statutory approval.

6.3 RISK RANKING METHODOLOGY

In order to effectively and qualitatively rank risks at the current stage of design an appropriate Risk Assessment Matrix (RAM) has been adopted and shown in Figure 6.2.

The orange and green areas of Risk Assessment Matrix represent the range where the risks are tolerable if they are "As Low As Reasonably Practicable" (ALARP). This is achieved if the cost of risk reduction would greatly exceed the safety improvements gained. The greater the risk, the greater the disproportion between cost and benefit which is required before an improvement becomes impractical. The red shaded region represents the range where risks are intolerable and would be required to be reduced by design activities.

The risks identified during the HAZID study have been classified according to their severity to the operations (downtime), impact on personnel on-board (health/safety), impact to the environment and their likelihood of occurrence.

6.4 HAZARD IDENTIFICATION & HAZARD OPERABILITY (HAZID & HAZOP)

The HAZID/HAZOP study took place at Imabari's Marugame Headquarters in Japan. The team review was led by a Chairman; the remainder of the HAZID/HAZOP team comprised of specialists from Imabari's design development team, owner's representative, fuel gas system designers, compressor system manufacturers, LR's subject matter experts (SMEs) and other specialists/manufacturers.

The main objectives of the HAZID study are to:

- Identify potential hazards and Class approval issues associated with the various proposed gas supply and auxiliaries arrangements for incorporation of the ME-GI engines.
- Identify potential hazards and Class approval issues associated with the gas fuel piping distribution to the ME-GI engines.
- Identify potential hazards and Class approval issues associated with the proposed HP-FGSS arrangement within Cargo Machinery Room.
- Identify potential hazards and Class approval issues associated with the proposed position of the gas valve units (GVUs) and any issues related to HP natural boil-off gas (NBOG) management and the GCU proposed operability.
- Assess the adequacy of the proposed SOLAS provisions and identify any Flag compliance issues/requirements.
- Issue HAZID Report inclusive of work sheets identifying findings, recommendations and actions.

Each part of the proposed system/operations and the area(s) which would take place on-board the ship was reviewed in turn by the HAZID team, applying the guide words or considering potential scenarios, to identify potential hazards. Causes of the potential hazards with resultant consequences were identified and risks were ranked based on the methodology presented in Section 6.3.

Class requirements and design safeguards were recorded and the implementation of any future control and safety measures was noted. Recommendations were made with respect to any additional mitigation measures by detailed design to minimise risk levels.

The main objectives of the HAZOP study are to:

- Undertake a detailed examination of the proposed ME-GI Fuel gas supply system, on a line by line basis, in compliance with the formal HAZOP process.
- Assess the consequences of upset conditions including flow, temperature and pressure and identify potential hazards associated with all aspects of the operations including the following aspects:
  - HP-Compressor System
  - HP-Pump System
  - HP gas header to gas master valve and to GVU of ME-GI
  - HP relief system to GCU
- Assess the adequacy of the operating procedures to prevent or control the potential hazards and incorporate any additional measures to the Operating Manuals as appropriate.
- Issue HAZOP Report inclusive of work sheets identifying findings, recommendations and actions.

The HAZOP team reviewed the above aspects applying the parameters/deviations and considering scenarios, to identify potential hazards.

Causes of the potential hazards and resultant consequences were then identified, together with the implementation of any additional design, isolation, control and safety measures. Where necessary, recommendations were made with respect to changes in design and/or implementation of operation procedures to minimise risk levels.

6.5 SAFETY ACTION REGISTER (SAR)

In line with the requirements of a Formal Risk Assessment submission, all actions identified as a result of HAZID/HAZOP studies shall be incorporated in Safety Action Registers (SAR).

The purpose of such register is:

- To provide project engineering parties with formal requests for actions related to Risk Assessment findings.

- To assist understanding and facilitate agreement among parties on the scope of actions required and enable the reporting of these actions.
- To establish a formal audit procedure for Class Plan Approval actions required especially for issues not referenced within existing Rule requirement.

6.6 INCORPORATION OF THE RESULTS OF RISK ASSESSMENT INTO DETAILED DESIGN

Based on SAR of HAZID/HAZOP studies, revision and supporting studies were performed at detailed design stage.

One of the examples is application of duct system instead of double wall pipe system for HP-Gas Supply Line which is arranged from gas master valve in the cargo area to GVUs in engine room, which achieved equivalent safety while improving maintainability and accessibility.

The HP Gas Line duct from the gas master valve to engine casing is controlled at slightly above atmospheric pressure with nitrogen circulation, and the duct from engine casing to GVUs in engine room is mechanically ventilated with more than 30 times volume per hour.

Several HP-pipe rupture scenarios were investigated by use of CFD analysis performed by LR's Technical Investigations Department, and the effectiveness and integrity of the duct system was ensured by taking into account the pressure and temperature effects in the design. An example of the simulation study of high pressure pipe rupture is shown on Figure 6.3.

7. CONCLUSIVE REMARKS

A technologically advanced LNG tanker has been developed by Imabari Shipbuilding with the collaboration of Lloyd's Register in order to meet the demand of a new global LNG trade.

Some of the outstanding features of the design are dual fuel diesel engines directly driving twin propellers with low fuel consumption, flexible operation of fuel gas supply systems and the reliability and safety considerations.

This paper highlights these design features and the measures adopted to ensure stringent standards in safety, maintainability, performance and habitability are met; including HAZID/HAZOP of high pressure fuel gas supply system, strength & fatigue verification, and noise and vibration assessment of the hull and containment system.

When the first ship of the series is delivered in 2017/Q2, she will be amongst the largest capacity and most energy efficient LNG tankers able to transit the Panama Canal, and the largest membrane LNG tanker constructed in Japan up to that point, setting a new paradigm for the LNG shipping industry.

8. ACKNOWLEDGEMENTS

The authors wish to thank colleagues in Imabari Shipbuilding Company Limited and Lloyd's Register Group Limited for their support in the preparation of this paper.

We wish to especially thank the following persons: Mr Hitoshi Fujita of Imabari for his sponsorship to publish this paper, Mr Kazuhiko Takeda of Lloyd's Register for his technical guidance throughout the project, Mr Sotirios Yiannakis for his contribution to the part on vibration analysis and Mr Benjamin Trimoreau for the part on noise analysis.

GTT's permission for the use of contents in Figure 2.2 is gratefully acknowledged.

DISCLAIMER

The views expressed in this paper belong solely to the authors and do not necessarily represent those of either Imabari Shipbuilding Company Limited or Lloyd's Register Group Limited.

9. REFERENCES

1. 'BP Statistical Review of World Energy June 2016', 2016, *BP p.l.c.*

2. 'SHIPRIGHT Design and Construction, Additional Design Procedures, Sloshing Assessment Guidance Document for Membrane Tank LNG Operations, Version 2.0 May 2009', *Lloyd's Register.*

3. 'Rules and Regulations for the Construction and Classification of Ships for the Carriage of Liquefied Gases in Bulk, August 2013', *Lloyd's Register.*

4. 'SHIPRIGHT Structural Design Assessment (SDA), Primary Structure of Membrane Tank LNG Ships Guidance on Direct Assessment, October 2013', *Lloyd's Register.*

5. 'SHIPRIGHT Design and Construction, Fatigue Design Assessment - Application and Notations June 2015', *Lloyd's Register.*

6. 'SHIPRIGHT Design and Construction, Fatigue Design Assessment, Level 3 Procedure: Guidance on direct calculations, March 2009', *Lloyd's Register.*

7. 'SHIPRIGHT Design and Construction, Construction Monitoring, May 2004', *Lloyd's Register.*

8. VIOLETTE, F.; SHENOI, A., 'On the Fatigue Performance Prediction of Ship Structural Details', *Transactions of RINA*, 1998.

9. 'SHIPRIGHT Structural Design Assessment Procedure – Sloshing Loads and Scantling Assessment, May 2004', *Lloyd's Register.*

10. 'Guidance Notes General Overview of Ship Structural Vibration Problems, September 2015', *Lloyd's Register.*

11. 'Ship Vibration and Noise, Guidance Notes, July 2006, Revision 2.1', *Lloyd's Register.*

12. 'SHIPRIGHT Design and Construction, Additional Design Procedures, Procedure for Analysis of Pump Tower and Pump Tower Base, September 2008', *Lloyd's Register.*

13. HOLLAND, C. G.; WONG, S. F., 'Noise Prediction and Correlation with Full Scale Measurements in Ships', *Transactions of IMarE*, Vol 107. Part 3, pp 195-207.

14. 'SHIPRIGHT Design and Construction, Additional Design Procedures, Risk Based Designs (RBD), May 2016', *Lloyd's Register.*

10. AUTHORS BIOGRAPHY

Hisatsugu Inoue holds the position of Engineering Manager of Ship Planning Team at Imabari Shipbuilding's Marugame Headquarters.

He is responsible for the initial design of cargo handling system and fuel gas supply system for LNG tankers.

He joined Imabari Shipbuilding Co., Ltd. in 1999 at the Hull Outfitting Design Team where he had been in charge of initial design of cargo handling system for Imabari's first LNG tanker project delivered from 2008-2010.

He is a graduate from the Marine System Engineering Department of Osaka Prefecture University.

Taichi Nakatani holds the position of Assistant Manager of Ship Planning Team at Imabari Shipbuilding's Marugame Headquarters.

He is responsible for the initial design of machinery for LNG tankers.

He joined Imabari Shipbuilding Co., Ltd. in 2006 at the Machinery Outfitting design team.

He is a graduate from the Aerospace Engineering Department of Nihon University.

Kazuaki Yuasa holds the position of General Manager for Japan, Lloyd's Register Marine & Offshore. He is responsible for all marine business in Japan area. He recently retired from LR in June 2016.

He joined Mitsubishi Heavy Industries Ltd. (MHI) in 1977 at the Ship Engineering Department in Tokyo, transferring to Nagasaki Shipyard at the Ship Designing Department, and returned to Tokyo in 1984 at the Ship and Ocean Engineering Division where he was active in the development of LNG & LPG carriers, chemical tankers, product carriers and oil tankers, engaging in Australian NWS, Taiwan-Indonesia, Abu Dhabi & Qatar projects.

He transferred to MHI Europe Ltd. in London from 1993 to 1996 as General Manager of Ship Sales and Technical Representative, and returned to Tokyo as LNG & LPG Carrier Project Manager in the Ship and Ocean Engineering Department, where he was involved in various LNGC projects such as Brunei, Shell G-class, MISC, TEPCO, Snohvit and Sakhalin projects.

He joined Lloyd's Register (LR) in 2007 as Global Gas Ship Business Manager who was responsible for promoting and growing LR's gas carrier business, both in Japan and globally, and promoted to General Manager for Japan in 2009.

He is a member of the Board of Councillors of the Marine Disaster Prevention Centre since 2013 and a Visiting Professor of Yokohama National University since 2014.

He is a graduate of the University of Tokyo (B.Sc and M.S.) majoring in Naval Architecture.

Henry Chung holds the position of Principal Specialist and Technical Consultancy Manager at LR's Yokohama Technical Support Office in Japan.

He joined LR in 1996 at the Technical Planning and Development Department in London where he was active in the development of Rules and Procedures for ships, in particular the *SHIPRIGHT* procedures. He has been based at the Yokohama Office since 2003 primarily engaged in the design appraisal of tankers and gas ships structures, progressing to his present position in 2010.

Prior to joining LR, he worked as a naval architect at Martec Limited in Halifax, Canada, now LR's Applied Technology Group.

He is a graduate of the University of Southampton (B.Sc.) and the University of California Berkeley (M.S.) majoring in Naval Architecture and Offshore Engineering. He is a registered professional engineer in Canada and a member of RINA.

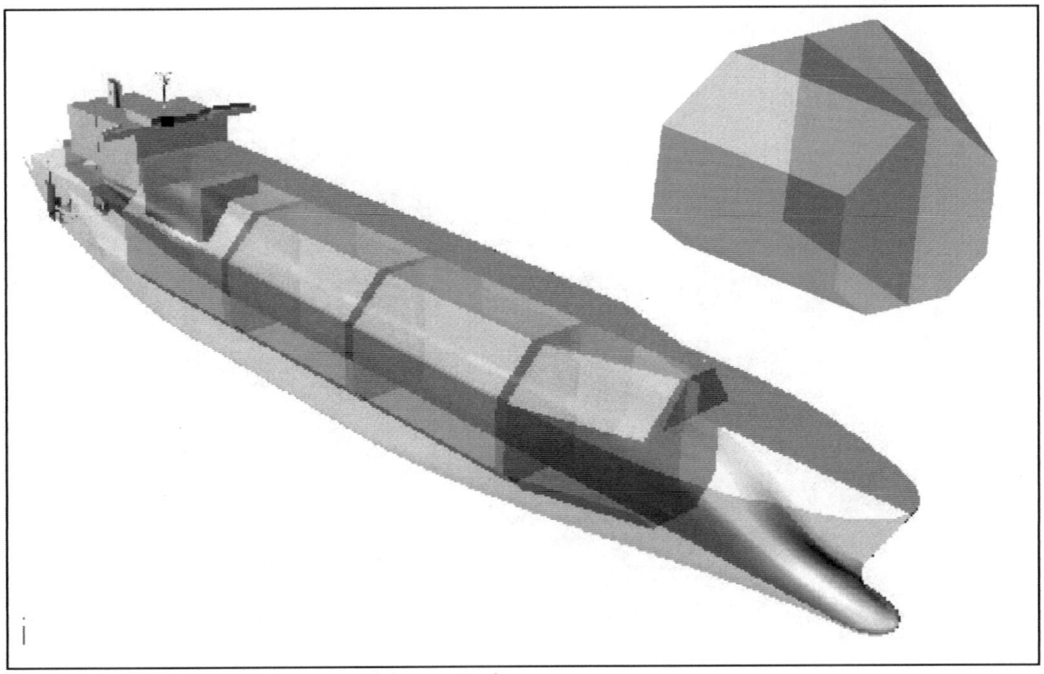

Figure 2.1: Configuration of Trapezoidal No.1 Tank

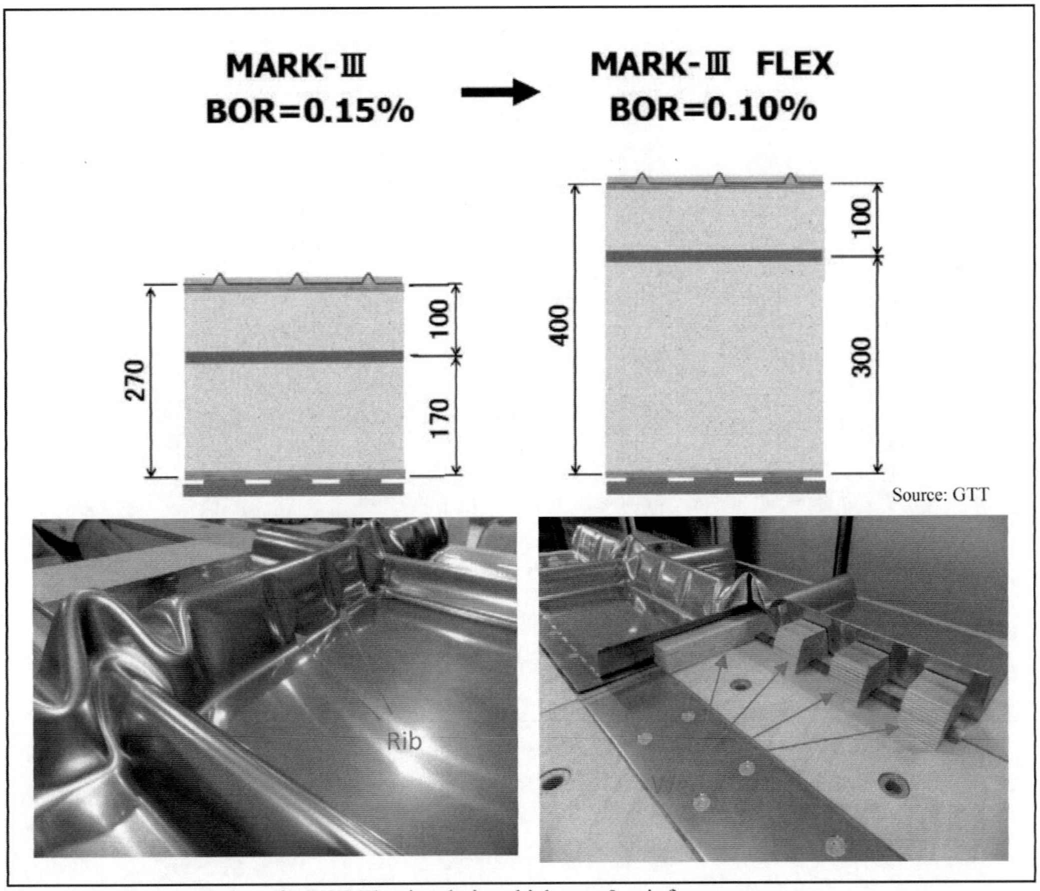

Figure 2.2: Characteristics of MKIII-Flex insulation thickness & reinforcements

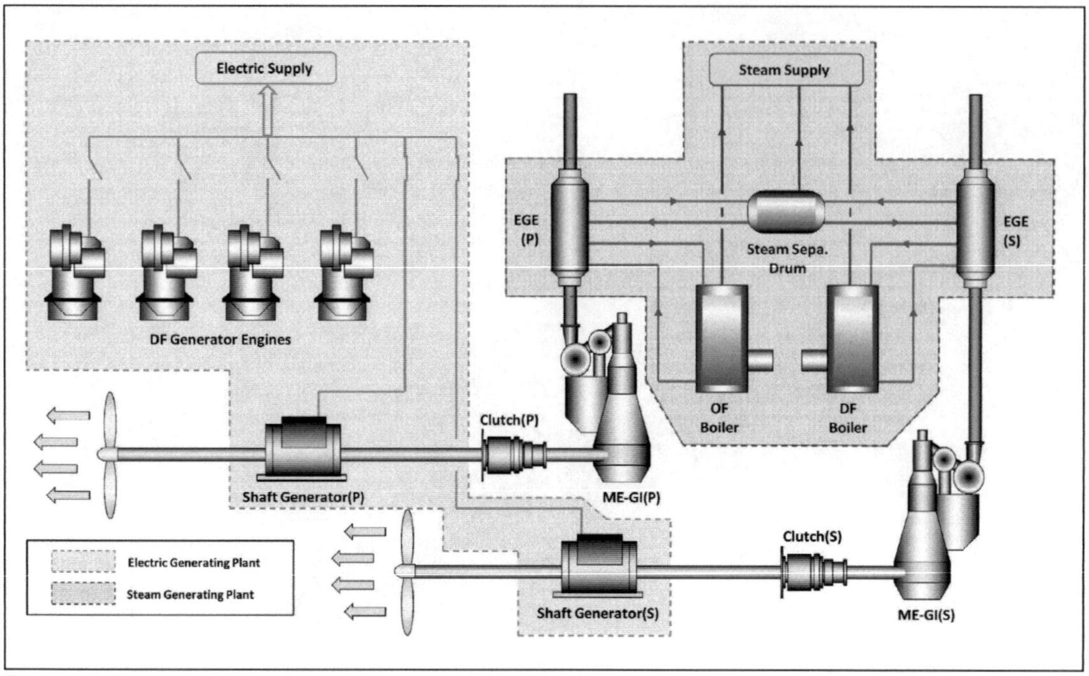

Figure 2.3: Propulsion, Electric and Steam generating Plant

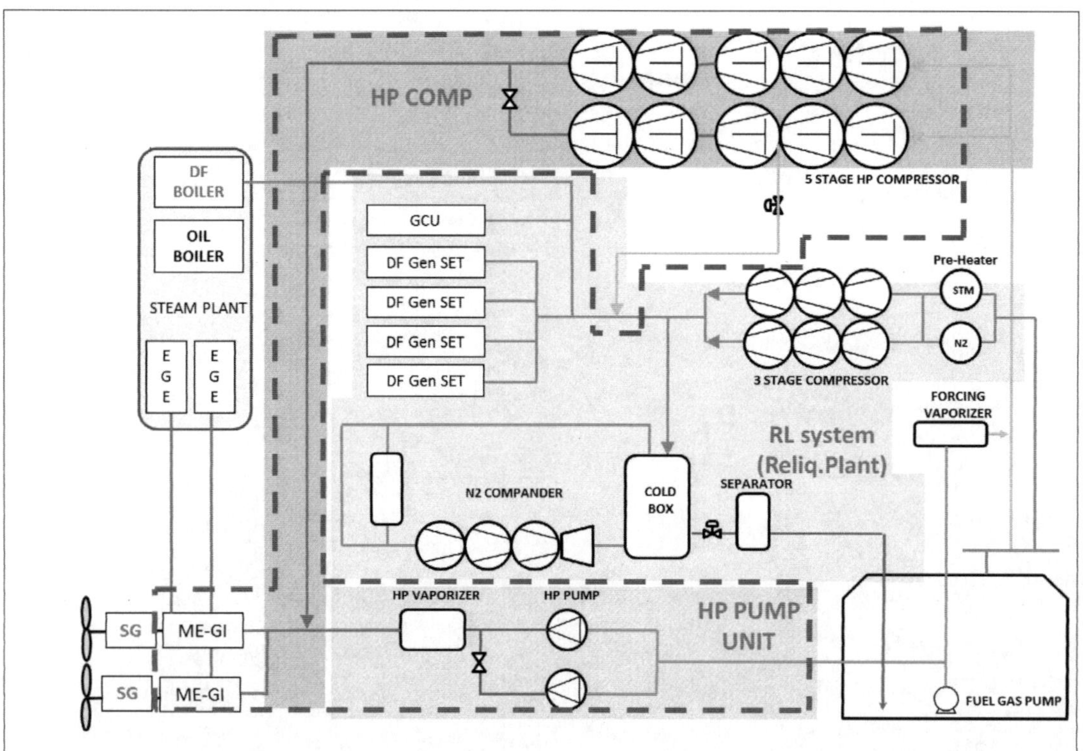

Figure 2.4: Fuel Gas Supply System and BOG handling system

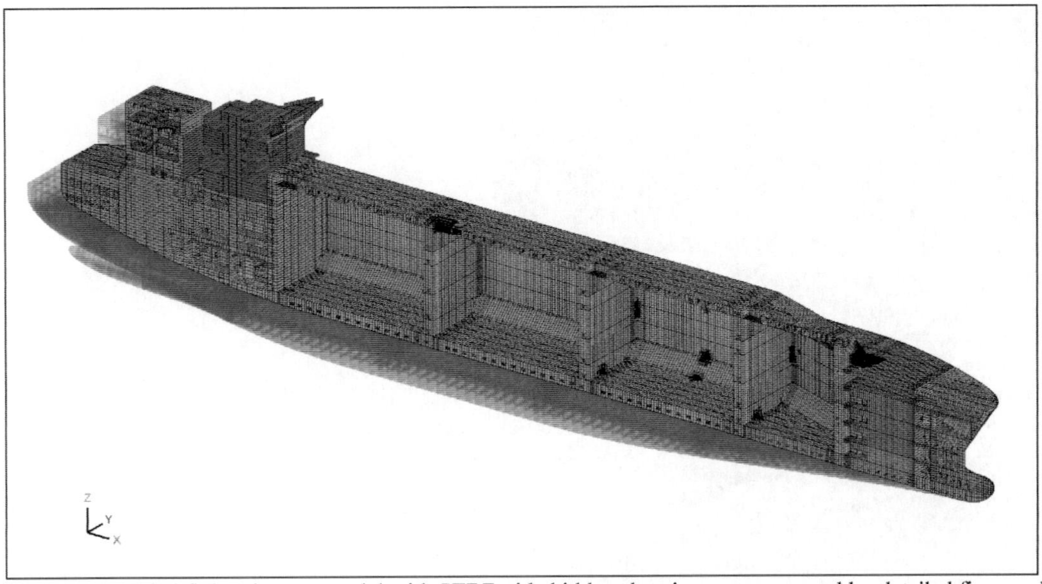

Figure 4.1: Global finite element model with STBD side hidden showing areas assessed by detailed fine mesh.

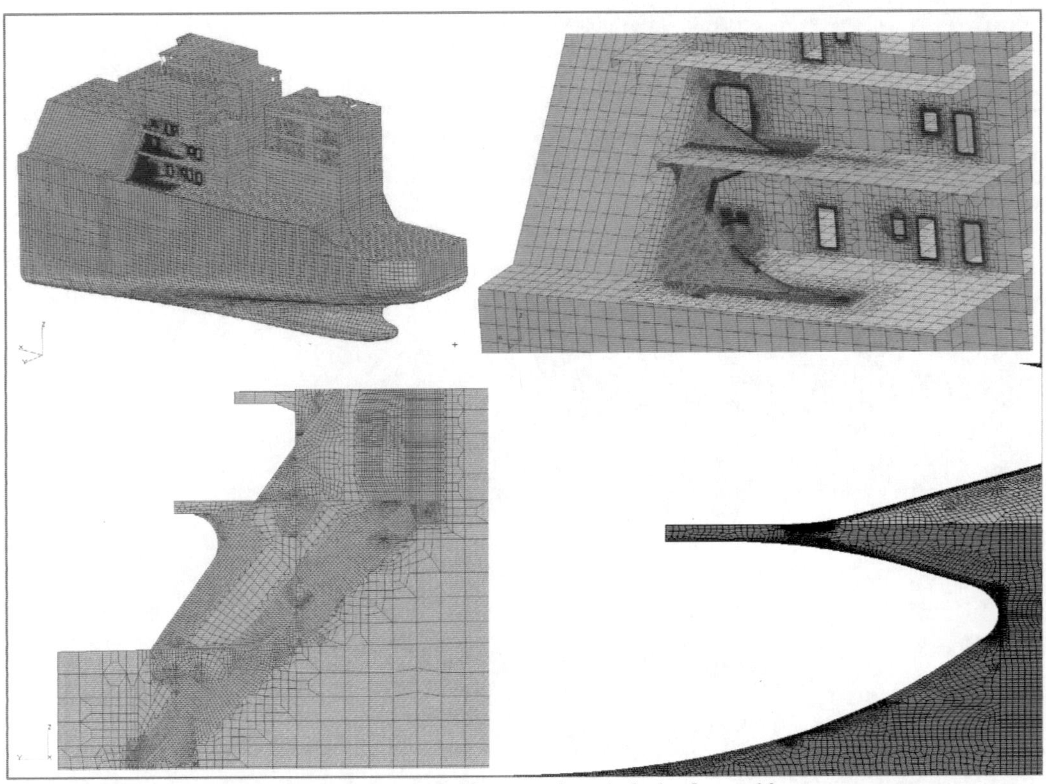

Figure 4.2: Stress and fatigue assessment model in way of trunk deck aft scarphing arrangement

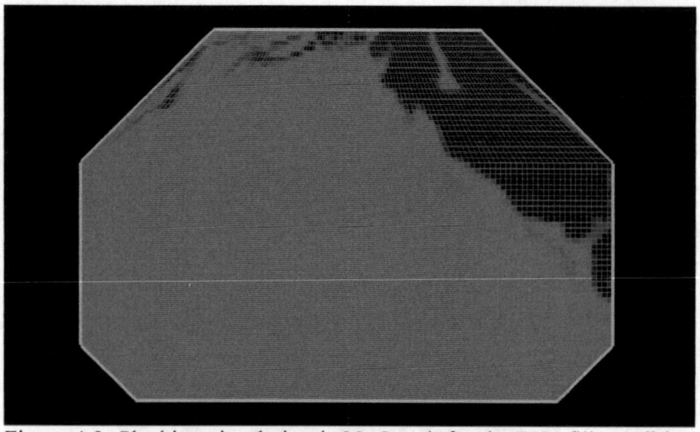

Figure 4.3: Sloshing simulation in No.2 tank for the 70% fill condition under transverse plane motions.

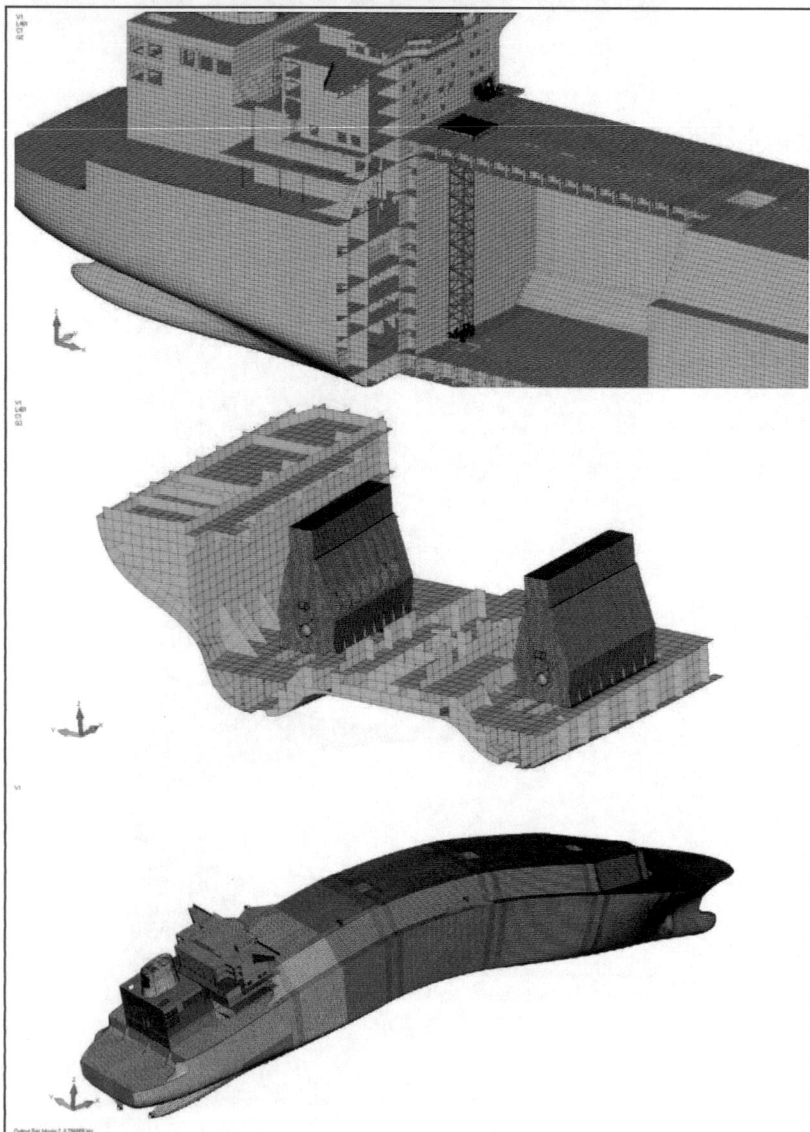

Figure 5.1: Modelled details in vibration analysis and first non-rigid body natural mode in loaded condition.

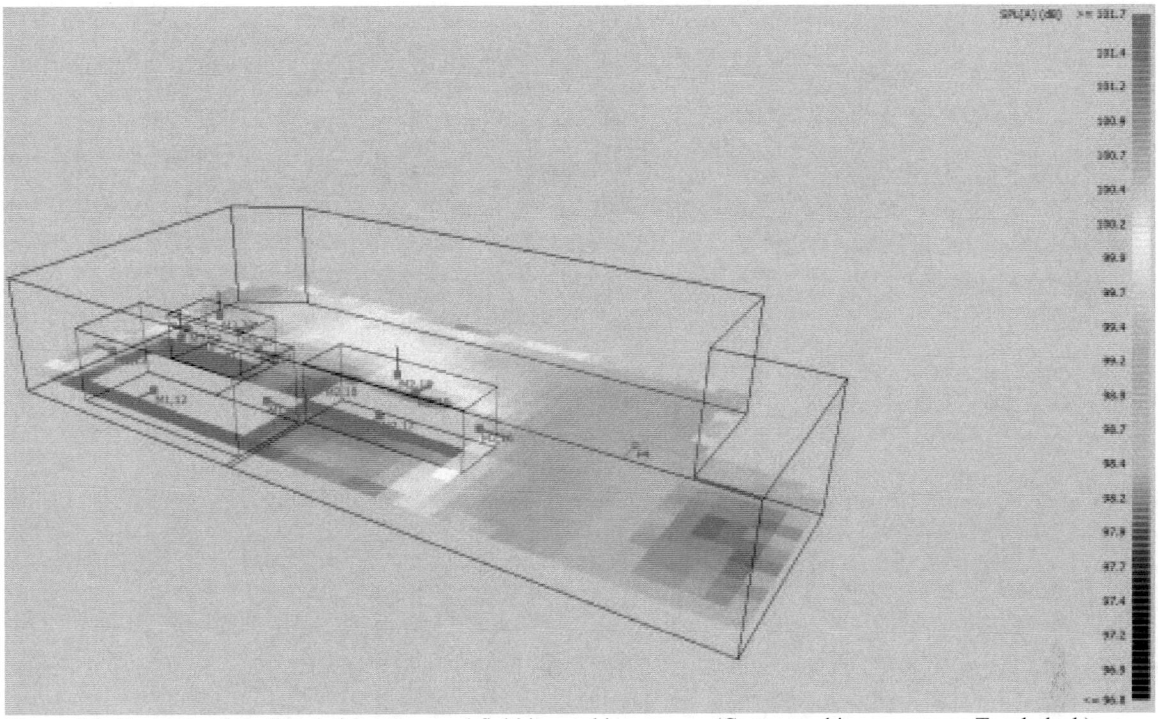

Figure 5.2: Example of predicted airborne sound field in machinery areas (Cargo machinery room on Trunk deck)

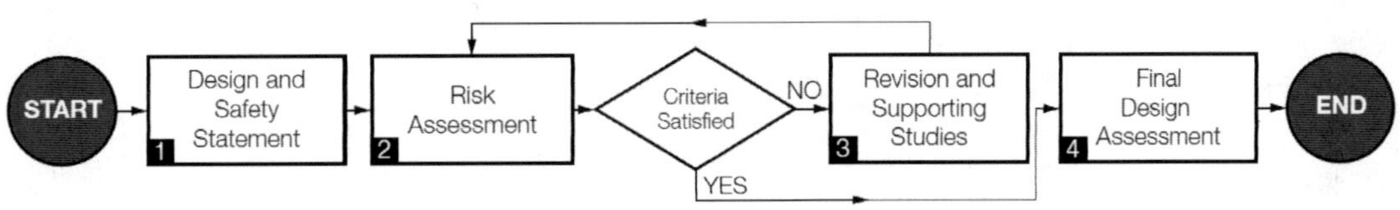

Figure 6.1: Risk Assessment Process

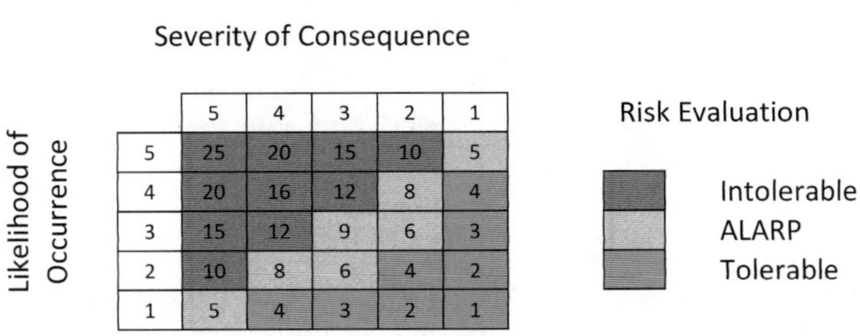

Figure 6.2: Risk Assessment Matrix (RAM)

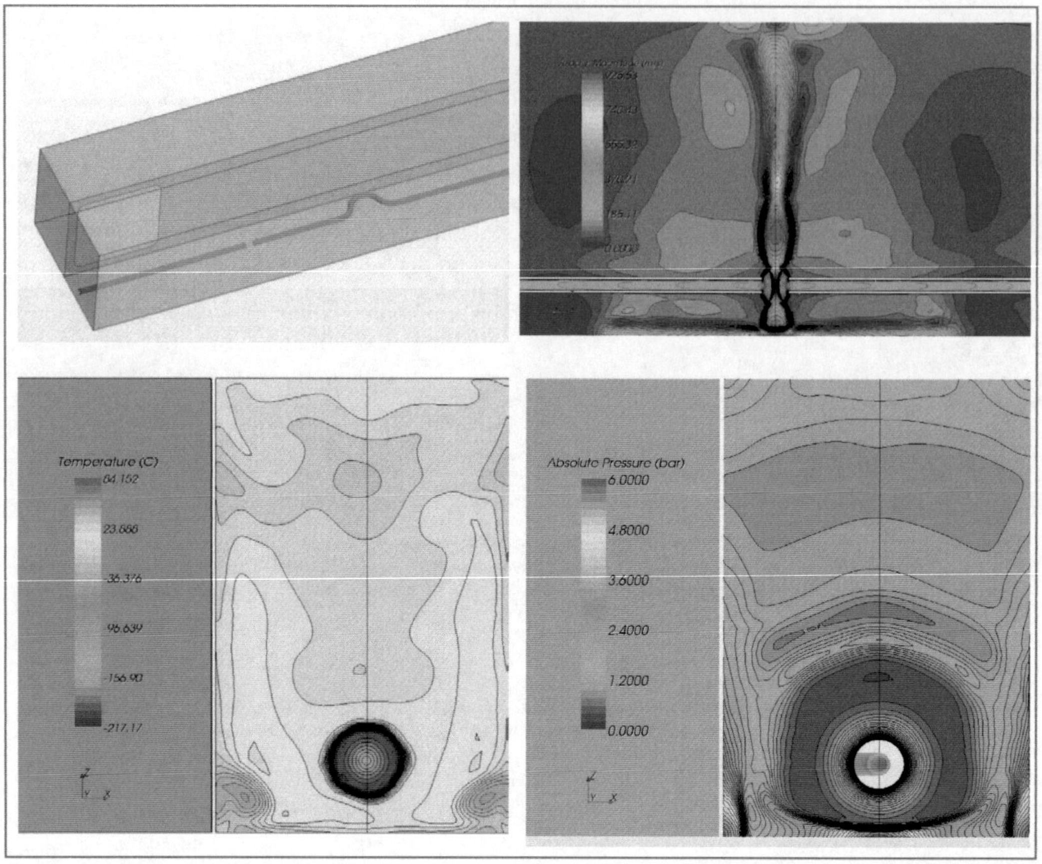

Figure 6.3: Example CFD simulation of high pressure fuel gas pipe rupture showing (clockwise from top left) location of rupture, velocity distribution, pressure distribution and temperature distribution at t=0.6 milliseconds from rupture.

SAFETY AND INNOVATION FOR LNG CARRIERS

Carlos GUERRERO, BUREAU VERITAS, FRANCE

SUMMARY

The LNG industry has developed safely and steadily. But the pace is speeding up now and flexibility and innovation have become a must. In this challenging market classification societies play a very significant role in ensuring the safety, feasibility and compliance of different solutions developed by shipyards, designers and equipment manufacturers.

New containment systems and new ship designs are under continuous development and propulsion systems have evolved significantly with two-stroke dual fuel engines a clear new trend.

NOMENCLATURE

NSR	Northern Sea Route
FSRU	Floating Storage and Regasification Unit
FSU	Floating Storage Unit
IGC Code	International Code for the Construction and Equipment of Ships Carrying Liquefied Gases in Bulk
IMO	International Maritime Organization
BOG	Boil Off Gas
BOR	Boil Off Rate
AiP	Approval in Principle
DA	Design Approval
GASA	General Approval for Ship Application
FA	Final Approval
NR.467	Bureau Veritas Rules for the classification of steel ships
JIP	Joint Industrial Project
PVC	Polyvinyl chloride
RPU	Reinforced polyurethane
R-PUF	Reinforced polyurethane foam
GW	Glass wool
PU	Polyurethane
DFM	Dual fuel mechanical
Reliq	Reliquefaction
DFDE	Dual fuel diesel electric
TFDE	Triple fuel diesel eletric
MEGI	2-stroke dual fuel engine electronically controlled with gas injection at high pressure
XDF	2-stroke dual fuel engine low pressure injection
EGR	Exhaust gas recycling
LGI	Liquid gas injection
DF	Dual fuel
UST	Ultra steam turbine
STaGE	Steam turbine and gas engine
GCU	Gas combustion unit
LNG	Liquefied natural gas
NG	Natural gas
CCS	Cargo containment system
FBOG	Forced boil off gas
NBOG	Natural boil off gas
HPLS	High pressure low speed
LPLS	Low pressure low speed
LNGc	LNG carrier
FGSS	Fuel gas supply system
HP	High pressure
SOLAS	IMO International Convention for the Safety of Life at Sea
MARPOL	IMO International Convention for the Prevention of Pollution from Ships
GTT	Gaz Transport and Technigaz
OPEX	Operational expenses

1. INTRODUCTION

The first LNG carrier for commercial operations entered in service in October 1964 when the first LNG cargo was transported from an export LNG terminal in Algeria to the UK.

In the beginning, the LNG carrier design has been very innovative with the first solutions to contain the cryogenic cargo developed by US, and then UK and France.

This incipient industry became then a bit more conventional with the invention of the first membrane and type B containment systems by the Norwegians and French companies MOSS, Gaz transport and Technigaz respectively.

These first designs considered either independent spherical tanks or membrane type of prismatic shape. Later on a type B prismatic has also introduced by IHI in Japan. For some reasons the LNG shipping industry has developed sustainably focused on the membrane and spherical independent tanks until very recent dates.

In an LNGc the cargo tank insulation system is not perfect and vapours generated inside the cargo space have been usually burned on board to generate energy and to propel the ships.

The boil off rate which is related to the insulation properties and size of tank (i.e. relation between volume and area of insulation) has been a significant percentage of the cargo volume for long time and used as a fuel in a no so efficient propulsion systems.

Due to commercial aspects and taking into consideration the increase of cargo tank volume, the evolution of materials and technologies, the BOR became a very important parameter in the recent years. Therefore big efforts have been made to reduce the amount of vapours in the cargo tanks and the required amount of fuel for ship's propulsion.

2. THE LNG CARRIER DESIGN ON THE PRESENT DAYS

In the recent years, the traditional paths of the LNG transportation by sea have changed significantly.

The LNG carriers in the past were traditionally designed and built for a certain trade, i.e. from a designated export terminal to few regasification terminals. This A to B trade has now changed as a lot of new export and import terminals have been commissioned and long-time charter contracts are less and less in place.

Short term (1, 3 or 5 year contracts) or spot market is in the present days approximately a third of the total LNG trade by sea. This requires a more innovative design of ships and more flexible solutions to be competitive.

In addition, new markets with specific design requirements have been developed in the last years.

In addition, since 2005 FSRU's have become a flexible and relatively cheap solution to have a regasification terminal for countries with growing energy needs and willing to use clean fuels to produce power. FSRU are presently one of the fastest ways to become natural gas importer and having the advantage of flexibility on supply.

Besides, a potential marker of small scale LNG distribution, originally in Europe and the trend to use LNG as a clean fuel for ship's propulsion is now a reality even if the expected growing path has been affected by the current diesel and heavy fuel oil prices. As a consequence, small scale LNG carriers have been designed and built with a containment system traditionally used for small size LPG carriers, the cylindrical shaped Type C tank. However, some of these small LNG ships have been designed as multipurpose gas carriers, i.e. able to transport also liquefied ethane, liquefied ethylene and liquefied petroleum gas in order to be fully flexible and able to change cargo from time to time depending on the charter contract.

All in all, these significant changes have led to innovative designs which in any event must be assessed by the classification societies and international and national organization such as IMO.

3. CCS and PROPULSION SYSTEMS. THE ROLE OF THE CLASSIFICATION SOCIETY

It is known that the classification society role is to assure that the ship is designed, built and maintained within certain rules and regulations.

The classification society has a very important responsibility nowadays in this innovative ship segment as to certify that the new designs are in accordance with the regulations and do not compromise the safety and the environment.

Classification societies are also acting on behalf of the national authorities. For ships involved in international trade, the IGC code is the most important regulation in addition to SOLAS and MARPOL all regulations issued by the IMO.

The IGC code covers specific safety aspects for gas carriers which are not addressed by the other international regulations.

In addition, the classification societies include the IGC code and other regulations related to ship' structure among others issues in their classification rules.

As previously mentioned, originally membrane containment systems and type B spherical type were widely used for LNG carriers.

The chapter 4 of the IGC (and the same chapter of the recently revised IGC entered into force in 2016) address how to design and built a containment system to transport liquefied gas. In particular this chapter is "*to ensure the safe containment of cargo under all design and operating conditions having regard to the nature of the cargo carried*".

Modern CCS include the latest versions of the membrane systems developed by GTT, type B MOSS and IHI (spherical and prismatic shape) and type C (cylindrical shape). More recently a type A tank has been also developed by the Norwegian company LNG New Technologies.

For safety reasons a reliable system which contains the LNG and avoid leaks into the ship structure is required. In addition and following latest industry trends a good system is a system which also has a reduced BOR and then prevents the formation of excessive vapours in the tank. It is highlighted that venting of natural gas is not permitted by the IGC code.

Since there are many new systems developed or under different stages of development in the recent years, the classification societies have also developed a 3 tier approval process to satisfy the industry needs.

In particular, an early stage approval for a containment system is offered by Bureau Veritas when the technology is not fully developed or just in case the developer is no prepared to carry on a full assessment for any commercial reason.

This early stage approval is called AiP (approval in principle) and is mainly offered to confirm that there are now showstoppers and to identify the important issues which may lead to a non-compliant system. The advantage of the AiP is the relatively low cost and short time required to prepare a report and issue a certificate. No specific list of drawings or documents is required but a minimum amount of information related to the containment system is desirable as the future development of the system may be impacted by the outcome of this approval.

A DA is required (GASA in other class societies terminology) when the system is to be fully approved per class rules and/or IMO IGC code regulation. Therefore, all aspects mentioned in the regulations must be assessed. More detailed information has to be submitted to satisfy the classification society surveyors in the task to assess the system against the required regulations.

The final approval process (FA) takes places when a system is DA and is installed on board of a ship. In Bureau Veritas view this FA does not require additional assessment for the CCS as the DA covers suitably the needs for a full approval of the system. However, the final test and acceptance of the tank during the commissioning phase is indeed required to fully approve the system in accordance to the regulations.

Taking into consideration the large amount of new systems developed over the last years, some of the systems have been only approved in principle as the developer is waiting the time when a potential application for a ship design comes.

Efforts by the containment system developers have been made in order to provide solutions in compliance with the regulations but providing a competitive BOR. It is necessary to mention that in conventional CCS of atmospheric type the cargo vapours must be burned but cannot be vented.

Burning the boil off gas is a handicap for the commercial operation as the charterers want to reduce the OPEX with regards to cargo vapour losses. Therefore, BOG management becomes a key issue for modern and efficient LNG carriers.

In addition, for some specific applications such as the FSRU or FSU, a CCS which grants no sloshing deformations as to avoid potential leaks at partial loading conditions is required. Therefore containment systems have been reinforced, in particular membrane systems, offering a reliable solution for partial loading conditions applications.

In the assessment of cargo containment systems and in particular for some applications and some specific systems, sloshing studies are very relevant.

Conventional systems used for LNG carriers with membrane technologies had some prohibited tank filling levels for the CCS. Levels from 10% up to 70% on the tank height are not allowed in a normal membrane containment system for example. The reason for this is basically the sloshing forces created in the liquid during navigation which are normally maximum in the range of 25% to 30% of the tank height.

It is worth mentioning that during the normal operation of an LNG carrier, a membrane tank is traditionally filled up to 98% when the ship is loaded at the export terminal and after offloading in the regasification terminal a little cargo remains normally in the tanks, which is called heel, in order to keep the CCS cool and to have enough fuel for the return trip in ballast.

However, considering the flexible LNG trade nowadays, the ships should be designed and prepared to transport cargo and preferably without barred tank filling, in particular if the ship is dedicated to FSRU or FSU applications.

In fact for LNG carriers a full flexibility in terms of LNG volume able to be transported may be wished because of the different trade patterns. Consequently, charterers and shipowners prefer to have the ability to cargo any range of LNG volume in different configurations within the CCS manufacturer recommendations mentioned above (out of the prohibited range between 10 and 70 per cent).

Regarding boil off, conventional membrane systems for example, developed since 1995guaranteed a 0,15% per tank volume per day of BOG which matched with the traditional propulsion systems based on steam boilers and turbines. Nevertheless, the new CCS evolutions have significantly reduced the BOR to 0,10% and are targeting levels even below with improved insulation materials.

These reduced BOR levels have been requested by the charterers to the shipowners aiming to reduce the OPEX of the LNG carrier.

The BOR levels must be also in line with the present technologies for gas fuel applications on board of ships.

IMO IGC Code covers the use of natural gas as fuel (vapours of the cargo tanks) for power production on board the LNG carriers.

The first systems on board of LNG carriers required a larger amount of vapours for the production of power. A close loop steam-condensate was the preferred design for the conventional ships as the BOR matched the consumption of natural gas in the steam boilers to drive steam turbines for propulsion and electricity production. During low steaming, the boilers are still burning gas and steam vapours not used for propulsion or electricity production are condensed again in the so called steam dump. It is proven that the system efficiency is less than combustion engines but the technology and commercial requirements did not lead into a revolution in the propulsion concepts until 2004.

In 2004 a first main technology change was introduced in Chantier de l'Atlantique (France) for a 75.000 m3 LNG carrier. Wartsila, the shipyard and Gaz de France (currently ENGIE) teamed up for a new concept of LNG carrier. The ship was delivered in 2006 based on the DFDE concept, i.e. generator sets able to burn liquid fuel (marine diesel oil or heavy fuel oil) or gas in a four stroke Otto cycle combustion engine.

A DFDE concept is then usually based on four or five generator sets (equal in size or not) a main switchboard for electricity distribution to the consumers and one or two main propulsion drives to propel the ship (in single or twin screw configuration). Additionally a GCU is installed in the engine casing to burn the remaining cargo vapours either in low steaming or during manoeuvring, stay at port, etc, as the vapours can only be vented in case of emergency.

The DFDE concept was widely implemented after 2006 and has been until 2014 the preferred option for LNG carriers being still nowadays the workhorse of the LNG carrier industry.

Anyhow, more efficient propulsion alternatives have been developed in the last years offering lower boil off consumption and being much more competitive in the present spot and flexible LNG carriers market.

Two stroke engines, which have been used to propel ships since the beginning of the modern era of shipping, were not able to burn gas or a mix of gas and liquid fuels for marine applications. In 2012, the commercialization of the new MEGI engine by MAN B&W introduced a competitive alternative which was selected for some of the newbuildings contracted in the last 3 years.

This type of propulsion is basically the same as the mechanically driven solution applied for the Qatari fleet of LNG carriers but the main difference is the introduction of the DF concept on the 2 stroke engine segment. In this type of ships a GCU unit is installed in any case for the reason mentioned above. Because of the low gas consumption, the DF MEGI application gives the designer some more room to think about partial or even full reliquefaction systems as the engine is running on a diesel cycle and therefore the consumption of natural gas is less than with the four stroke (Otto cycle). On the other hand, because of the diesel cycle requirements, the pilot fuel (liquid fuel to ignite the gas-air mix) is however higher than the four stroke solution.

A more recent technology has been also developed by Wartsila. The LPLS engine called XDF is basically a 2 stroke engine with gas injection at low pressure and running on Otto cycle. The engine room layout is similar to the MAN MEGI solution. Two main engines driving each one propeller and four stroke generator sets of smaller size for electricity consumers is the basic machinery arrangement, so a twin screw configuration is considered. A GCU is also installed in the engine room casing. The need for full or partial reliquefaction will be dictated by the operational profile of the ship. The total natural gas fuel used for propulsion is less than with the four stroke DF engine but a bit more than the MEGI solution. However, the required pilot fuel is less than the necessary amount for a MEGI engine making the total fuel consumption similar for both two stroke solutions, either high or low pressure.

On the other hand, significant researches have been made by Japanese shipyards involved in the construction of LNG carriers to promote new steam turbine systems.

The new steam generation systems are also aimed to reduce BOG consumption as the BOR of the MOSS type CCS is in the range of 0,10% per volume per day.

The first innovative system developed recently in Japan is based on the use of waste heat for reheating purposes giving some additional efficiency to the overall system.

Another system developed a bit later is a short of hybrid concept, using DF or gas engines and the waste heat recovered from these combustion engines is used for reheating purposes of the steam system. As a consequence, the electricity is generated on board by means of gas engines and/or steam turbines.

As described above, different efficient propulsion systems are developed and approved for LNG carriers. Consequently, the need for reliquefaction plant depending on the operational profile and economic factors is a fact.

Full reliquefaction systems on board of LNG carriers were designed and installed in the Qatar fleet of Q-Max and Q-Flex ships. These ships had a very specific requirement to reliquefy all the amount of BOG to save the cargo and keep it cool at any time.

Reliquefaction systems are a challenge. The design of this equipment has to take into consideration motions, such as rolling, pitching, etc which have never been design considerations for LNG plant on shore. In addition, reliquefaction systems are complex and expensive systems based on cryogenic moving components such as piston or centrifugal compressors.

For the very new two stroke DF engine solutions above described, considering also that the normal operation speed might be well below the design speed of 19,5 knots, technical solutions have been developed with different compressor type and connected to the gas supply system or independent.

As of today, only two of these 2 stroke mechanically propelled LNG carriers equipped with partial reliquefaction systems are in service.

Having in mind all the above, some classification societies have developed specific tools to guide the different actors during the design of a well optimized LNG carrier.

In particular, Bureau Veritas presently offers different alternatives such as:

- CFD calculations over range of speeds and in both laden and ballast condition to optimize the hull form, appendices and propeller(s).

- Ship Energy Efficiency Calculation and Analysis Tool (SEECAT software) which simulates entire energy cycle including domestic and cargo loads, GCU, Forced Boil Off generation for different power plant configurations and realistic voyages to optimise the choice of main propulsion, electrical generation systems and other big consumers on board, such as reliquefaction systems.

Two fully detailed joint projects have been developed in relation with LNG carriers optimization in the last couple of years. The first project was developed in cooperation with TOTAL, SHI for the optimization of a 180.000 m3 LNG carrier design. The second has been concluded in its first part and has been led by BG with the active participation of Bureau Veritas and GTT among other players.

Based on the outcome of the first project, two LNG carriers are presently being built in SHI (South Korea) for SK shipping. The ships are the first ever LNG carriers equipped with XDF engines.

4. CONCLUSIONS

The LNG shipping market is becoming highly competitive and fluid, which demands agility and technical innovation from all players.

Same agility is requested to the entities involved in the assessment and certification of new systems.

Consequently there is an additional effort requested to the classification societies which are responding to the clients by granting the necessary approvals aiming to preserve the excellent safety records of the LNG carrier industry.

5. AUTHORS BIOGRAPHY

Mr. **Carlos Guerrero** joined Bureau Veritas in September 2012 and since then he is involved in the business development of Oil/Chemical Tankers and Gas Carriers working in the Head Office in Paris. His main role is the field of oil and gas, following closely different subjects such as the use of LNG as a fuel, innovation in gas carrier designs, new gas containment systems and propulsion systems, risks studies, etc.

Before joining BV, Mr. Guerrero has been largely involved in shipbuilding activities in several Spanish shipyards, working in the production department, technical office, and finally in the commercial department. Carlos Guerrero has also ship operational experience because of his previous position as commercial manager of Union Naval Valencia, S.A. shipyard belonging to Spanish shipping company Boluda Group.

Mr. Guerrero is M.Sc. Naval Architect and Oceanic Engineering by the Universidad Politécnica de Madrid (Spain), where he finalised the thesis in March 1998 and is member of Spanish Official College of Naval Architects.

LNG AS FUEL CONTAINER VESSELS WITH MEMBRANE SYSTEMS

T Clément, P Charbonnier and M O Quéméré, GTT, France

SUMMARY

In order to respond to the more stringent regulations of the Emission Control Areas (ECA), liquefied natural gas (LNG) as a fuel is recognized as a valuable and more environment friendly alternative to heavy fuel oil (HFO). The membrane technologies developed by GTT have proven for decades their efficiency for large LNG transportation and storage. This paper deals with the solutions studied to implement a membrane LNG fuel tank in the case of large container vessels. Since those ships will operate in highly busy harbours or routes, the risk associated to ship collision is addressed through a safety assessment based on the structural response of the containment system under inner hull deformation and supported by experience at sea. In addition, the ability of membrane tanks to sustain sloshing loads for all filling levels during worldwide navigation is described. Finally, the practical industrial solutions to be considered during construction associated with the good integration of fuel tanks in the hull structure to maximize remaining space for containers are presented for different application cases – i.e. new-built or existing ships.

1. INTRODUCTION

Continuous increased requirements for lower particle emissions from exhaust gases drive the utilisation of HFO alternatives such as LNG. The latter fuel allows for lower sulphur oxide and nitrogen oxide emissions and less particles at a more competitive price compared to marine gas oil (MGO). However, the challenge with LNG is the inherent requirement to maintain a low enough temperature of -163°C to keep the natural gas under liquid phase at atmospheric pressure. As a consequence, dedicated fuel tank technologies are needed to limit the evaporation of the LNG and to ensure safe and efficient operations.

As of today, a fleet of approximately 80 LNG fuelled vessels, such as smaller car and passenger ferries, navigate primarily in the Norwegian ECA. Naturally, the next step is to utilize LNG as fuel for larger vessels where the available space for the tank is previously defined by the flexible integration of HFO tanks into the hull structure.

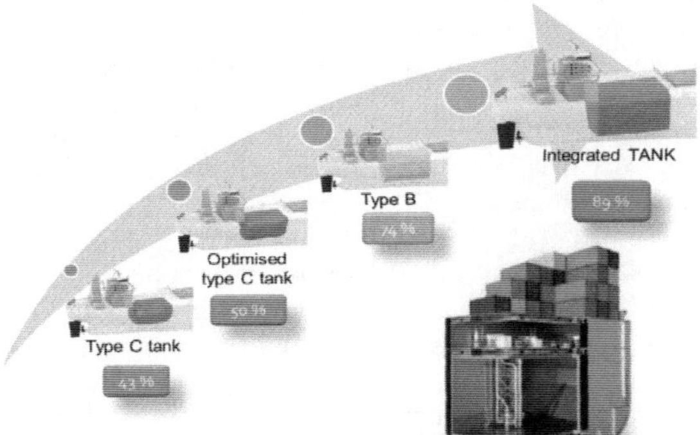

Figure 1: LNG tank space allocation in a given hold space and picture of a membrane fuel tank integrated in a container ship

As per IGF code, LNG tanks are classified through independent self-supporting tanks and integrated tanks. Independent tanks gathered the pressurized type C tanks without secondary barrier and the atmospheric type A and type B tanks with respectively full and partial secondary barrier. Membrane tanks, classified as integrated tanks, are atmospheric tanks with full secondary barrier.

Due to its modularity and its excellent integration inside a given hold space - see Figure 1 - , one key interest of considering a membrane tank is to maximise the LNG volume and consequently increase the ship autonomy. This is of high interest for the ship-owner to reduce the number of refuelling operations which are costly.

If the ship-owner prefers to fix the LNG volume, the membrane tank offers the opportunity to reduce the hold space for LNG fuel tank and increase the space for commercial cargo.

Through this paper, we will study the typical case of large container ship. In the first part, we will address the risk of collision and the consequences to be expected on a membrane tank. Then, the sloshing phenomena will be assessed for LNG fuel tank without filling limitation in worldwide navigation. Finally, the consideration of industrial models well adapted to the project execution for these vessels with LNG propulsion will be discussed.

2. DEDICATED COLLISION STUDY FOR CONTAINER SHIP EQUIPPED WITH GTT MEMBRANE TYPE LNG FUEL TANK

2.1 COMPLIANCE OF LNG FUELED CONTAINER VESSEL WITH IGF CODE

For the investigation presented in this part, a typical 18,000 TEU container ship is considered, see [1]. This ship type was originally designed to operate between Europe and Asia on heavy fuel oil, therefore fuel oil tanks were arranged below the deckhouse. The hull ship structure complies with the rules coming from DNV GL.

A large container ship can be either designed at newbuilding stage or retrofitted using LNG as a fuel. For the current study, the HFO tanks of 18,000 TEU container ship are replaced by two LNG membrane tanks. Based on the retrofit concept developed by GTT, the LNG tanks can be extended into the cargo hold behind the deckhouse, see Figure 2. The fuel gas supply system rooms are located on the upper part of the LNG membrane tank.

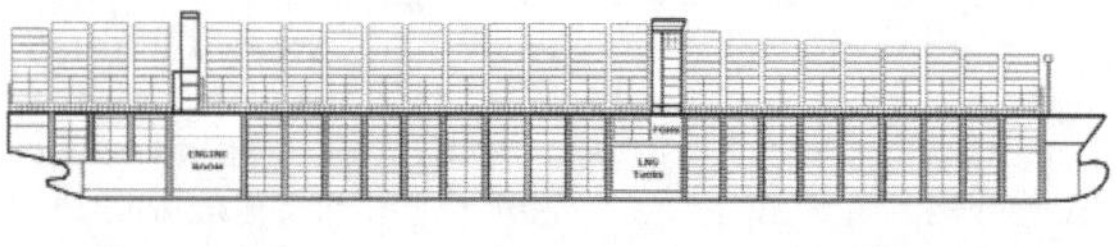

Figure 2: General arrangement of LNG fuelled 18,000 TEU container ship

In the longitudinal direction, the LNG membrane tanks are located between the support transverse bulkhead of the cargo hold and the forward transverse bulkhead of HFO tanks.

In the transverse direction, the LNG membrane tanks are located between longitudinal bulkheads and are symmetric with the centre-line, see Figure 3. Each tank has a volumetric capacity of 10,800 m³. The entire volume of LNG as fuel is enough for a complete round Europe-Asia-Europe.

From January 1st 2017, new rules (IGF code) for LNG fuelled ships will enter into force, see [2]. The IGF Code will regulate the arrangement, installation and operation of machinery and the entire system of the ship using LNG as fuel to ensure a safe management for the ship integrity and for the environment.

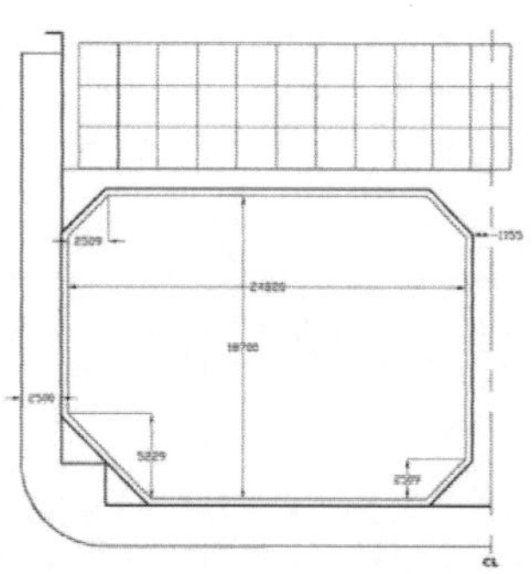

Figure 3: LNG fuel tank arrangement (transverse view)

According to the IGF Code, the LNG tanks should be located in such way that the probability for the tanks to be damaged in a collision or grounding event is reduced to a limit value. To define the limits for the boundary of the LNG tank, the IGF Code is proposing an empirical approach and a probabilistic approach. Based on the empirical approach, the LNG tank should be located at a minimum distance of B/5 or 11.5 m from the side shell, and at a minimum distance of B/15 or 2.0 m from the bottom shell. B represents the greatest moulded breadth of the ship at or below the deepest draught.

As an alternative to the empirical approach to define the limits for boundary of the LNG tank, the IGF code is also considering a probabilistic approach. This probabilistic approach is based on damage statistics collected from collision accidents as formulated in SOLAS regulation, see [3]. According to the probabilistic approach, a damage factor f_{CN} is calculated by using the formulations described in SOLAS regulation II-1/7 and it is composed of probability factor for longitudinal direction (f_l), a probability factor for transverse direction (f_t) and a probability factor for vertical direction (f_v). For container ships, the damage factor f_{CN} should be less than 0.04, which represent the penetration probability on a LNG tank in a collision event.

For the selected 18,000 TEU container ship, the design suggested by GTT has the distance between the outside insulation of the membrane tank to side shell equals to 2.50 m. It is the same distance to the bottom shell.

According to the IGF empirical approach, the required minimum distance between the LNG membrane tank and side shell is 11.5 m. Therefore, to identify the acceptable limits of the LNG tank on the 18,000 TEU container ship, the IGF probabilistic approach was applied. Using the calculation procedure described and IGF code, for the selected LNG tank arrangement, as shown in Figure 2 and Figure 3, it was calculated by DNV GL a damage factor f_{CN} equals to 0.02. In conclusion, the LNG tank arrangement of the 18,000 TEU container ship complies with all requirements of the IGF code.

2.2 PROBABILITY OF CONTAINER SHIP COLLISION

It is important to understand ship collisions and their possible consequences, such as loss of life, damage or total loss of ships and environmental pollution. Apart from preventing collisions in the first place, the objective of such research is to reduce the consequences of these collisions. The aim of this chapter is to assess the probability of container ship collision and it is based on Glaser [4]. The term ship collision refers solely to ship-to-ship collision.

As a first step to gather data on ship collisions, the public, non-commercial casualty database from IMO's GISIS was used. This search resulted in 95 collision cases in which a container ship was involved. The analysis of such a small number of cases does not lead to significant results. Therefore, DNV GL kindly provided an extensive dataset with collision cases from a recent project about Formal Safety Assessment (FSA) on container ships. This dataset was mainly comprised of data from two commercial marine databases LMIU and IHS Fairplay. The search criteria were maintained for the search in all three databases in order to obtain comparable results. The criteria included ship type "container ship", classed by "IACS society", built after 1981 and involved in a collision, reducing the cases from GISIS to a number of 89. Afterwards, the collision cases were compared and duplicate entries of cases that were listed in more than one database were deleted.

However, some datasets are incomplete and due to various reasons the striking ship is often not identified clearly. Therefore, additional information from various sources was collected in order to distinguish between struck and striking ship and to fill other missing entries. Reliable data were available from investigation reports published by national marine investigation branches and the European Maritime Safety Agency (EMSA). Further ship particulars relevant to the analysis were gathered from online register World Fleet Register by Clarkson's Research and GL legacy tool Ship Information System (SIS). Access to commercial World Fleet Register and SIS was also kindly provided by DNV GL. As a result, the following overview of collisions along the route connecting Europe and the Far East could be drawn, see Figure 4.

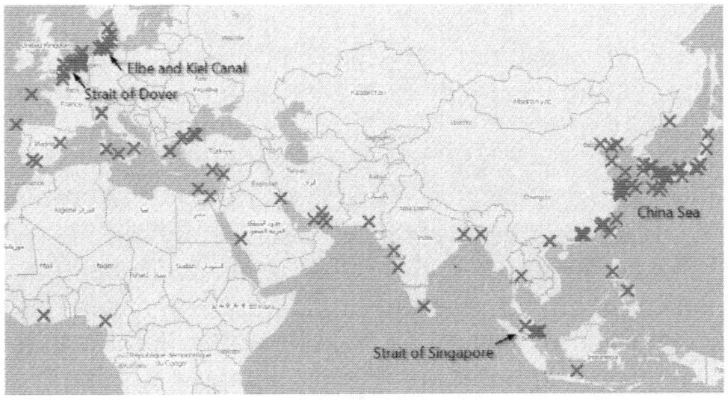

Figure 4: Locations of struck container ships in a collision from 1990 to 2014

Figure 5 shows the dependent probabilities leading to the probability of a rupture in the inner hull, in the parallel body of a container ship.

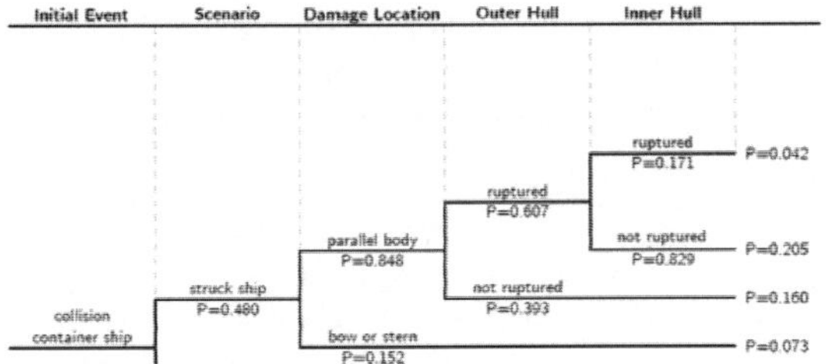

Figure 5: Probability of inner hull rupture

From the initial collision, the dependent probability of the inner hull being ruptured is p(inner hull rupture) = 0.042. The probability of inner hull rupture is calculated independently of the collision location or operating condition. The sample size of available data is not sufficient to break down the probabilities into further categories. Figure 5 shows the probability calculated for all locations combined. The overall probability of inner hull rupture for a container ship in a collision is then

p(collision with inner hull rupture) = p(collision) x p(inner hull rupture | collision)

The probability of inner hull rupture for container ships in a collision is 3.28×10^{-4} per ship year with the initial collision probability of p(collision) = 7.82×10^{-3}. This means that for a twelve year period on the worldwide existing fleet of container ships, the expected number of collisions for container ships is around 470 with about 20 of them resulting in inner hull rupture. The probability to collide especially at tank location is not taken into account which means that the probability to impact the fuel tank is even lower.

2.3 HIGH AND SAFE RETURN OF EXPERIENCE ON GTT MEMBRANE SYSTEMS

Since 1964, with the first LNG carrier built with a membrane technology, the shipping industry has accumulated a large experience; with an actual total fleet of 340 vessels shared between the two main technologies proposed: Mark and NO, the industry has learned a great deal in LNG maritime transportation. Considering the fleet along with 50 years of experience, the shipping industry has accumulated a feedback at sea which constitutes more than 15,000 years in cryogenic applications. During these years of navigation, no loss of tightness has been experienced by ships equipped with membrane tank.

In June 1979, the "El Paso Paul Kayser", while laden, sailing at 19 knots, ran head-on into a big rock. The impact energy was huge enough to crack the rock and the ship finished its course 100 m further on grounding on a second rock. The ballast compartments of the bottom outer hull were destroyed over a length of 180 m as shown in Figure 6 and the inner hull out of plane deformations reached nearly 30 cm [5]. This ship was equipped with NO82 technology. This experience has demonstrated the high flexibility of the membrane and the capacity of the containment to remain tight despite the severe deformation of the inner hull.

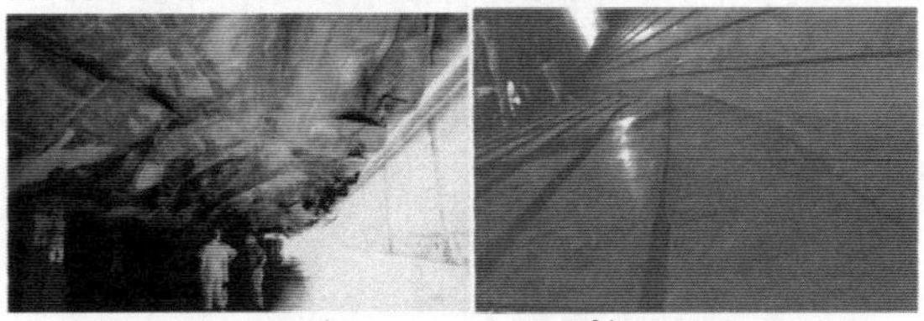

a) b)

Figure 6: El Paso Paul Kayser grounding: a) State of the ship bottom structure and b) Membrane deformation

In 2002, the Descartes, a ship equipped with Mark I technology experienced a large membrane deformation due to an over-pressure in the insulation space. An operational error on-board led to an over pressurization significantly above the atmospheric pressure. The primary membrane demonstrated its ability to sustain a deformation of 1.5 meters while remaining tight. An illustration of the over-pressurized membrane is provided below, see Figure 7.

Figure 7: Mark deformation due to over pressure

The industry feedback at sea has shown the ability of the membrane to sustain large deformations. To go further and to know the limits of the membrane in term of deformation, a full scale experiment has been realized.

2.4 FULL SCALE EXPERIMENT

2.4 (a) Mock-up and test rig

The mock-up design is defined by the limitations of the large scale collision test rig located at the institute for ship structural design and analysis of the Hambourg University of Technology (TUHH), see [6]. The large test frame consists of four 1 MN hydraulic cylinders mounted to two large crossbeams at the lower side and one large cross-member at the upper side (see Figure 8). A semi spherical bulbous bow like indenter with a diameter of 1.6 m is positioned at the cross-member above the center of the test mock-up. During the experiment, the indenter moves down and penetrates the mock-up. The indenter is designed to remain rigid during the experiment.

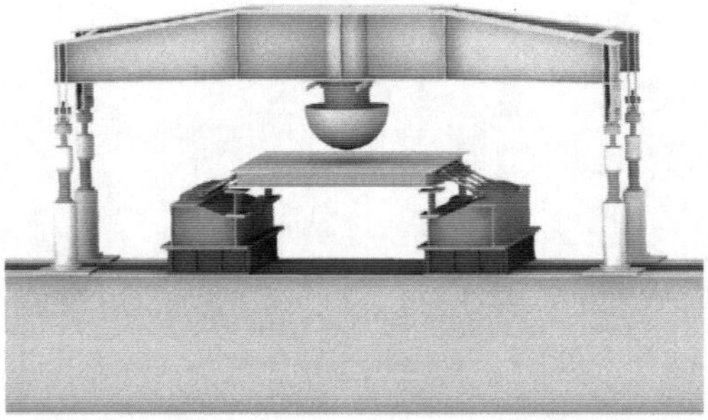

Figure 8: Full-scale 4 MN test frame at TUHH including the rigid indenter and mock-up

The mock-up represents a cut-out of the vessel's inner hull and the adjacent GTT Mark III membrane system. The hulls, as well as the inner barrier of the mock-up, are welded to a frame of 3.74 x 3.74 m which rests on the test rig. The steel plating of the mock-up representing the inner hull of the vessel is obtained in compliance with the deformation of the vessel's hull (global deformation) in the collision case. The deformations of the steel plating of the mock-up and the deformations of the corresponding 18,000 TEU container ship's inner hull are analysed using LS-DYNA. The striking bow is taken from a similar container vessel where the relatively sharp lines represent a more critical scenario as a result of the localized deformation leading to rupture of outer and inner hulls, see Figure 9.

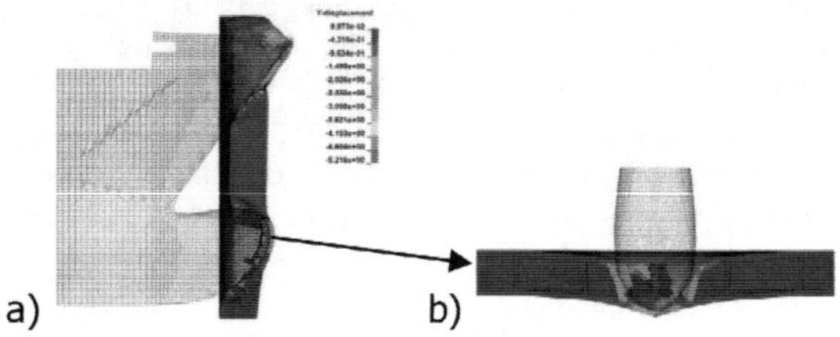

Figure 9: Simulated structural deformation prior to failure of the inner hull a) in y-z plane and b) x-y plane

The objective of the experiment was to deform the mock-up until failure of the primary membrane barrier occurs. The deflection is recorded using distance meters. In addition, strain gauges are placed on the upper cross-member as well as the frame and the inner barrier of the mock-up. The final mock-up assembly placed in the test rig is shown in Figure 8.

2.4 (b) Experimental results

The indentation proceeded continuously until the weld seam of the primary barrier and the frame collapsed. A further indentation would have caused a complete separation of frame and primary barrier. For that reason, indenting the mock-up further would not have provided any information on the load bearing capacity of the mock-up in general and of the primary barrier in particular. The attained deformation of the mock-up is displayed in Figure 10. The maximum attained displacement is approximately 1 m. The small and large corrugations started to unfold globally and unfold almost completely at the centre of the mock-up. However, the primary barrier at the centre remained tight as further dye penetrant testing revealed.

Considering the test set-up with a primary membrane sample of 3 m x 3 m surface, it would represent an equivalent penetration of the inner hull of about 8 m for a 24 m bulkhead span.

Figure 10: Deformation obtained on the primary membrane without loss of tightness at the end of the test

2.4 (c) Impact of cryogenic conditions

The test previously described has been conducted at ambient temperature. However to show that the conclusion would be similar at cryogenic temperature, we can use unfolding test results. These tests had been conducted at ambient and cryogenic temperatures. The objective was to unfold successively the large, then the small corrugation.

Results show that the membrane can be completely unfolded at both ambient and cryogenic temperatures without any crack initiation. Thanks to this last test, we can say that the conclusions of the collision test can be extrapolated at cryogenic temperature.

3. SLOSHING ASSESSMENT

3.1 CHOICE OF THE TANK ARRANGEMENT

One of the main advantages of membrane tanks is their adaptability to the vessel shape which leads to minimize the loss of commercial space. Two-row tank arrangement can be installed in the container ship to strengthen the ship structure and support heavy loads on the deck thanks to the longitudinal cofferdam. This arrangement can also offer more flexibility to the operator for fuel management or maintenance for instance. Moreover, in case LNG is the only fuel on board, the IGF code requires full redundancy of the gas system including two separate fuel tanks.

However a two-row arrangement is obviously less cost competitive as it requires more membrane surface and more equipment for fuel handling including two pump towers, two liquid and gas domes and a system to heat the longitudinal cofferdam. If a redundancy is not necessary, the best optimization between the LNG capacity and the space occupied is to implement a one-row tank arrangement.

A one-row arrangement induces a larger tank breadth, so sloshing is expected to be more severe compared to two-row tanks. A one-row tank arrangement will consequently require a strengthened fuel containment system (CS), but its global cost remains more attractive than two-row arrangement cost.

3.2 METHODOLOGY FOR SLOSHING ASSESSMENT

Sloshing is a non-deterministic phenomenon. Global ship motions are predictable but the magnitude, the location and the instant of a sloshing impact are rather unpredictable. Sloshing is hence studied through statistical approaches based on the description of the statistical pressure distributions and the frequency of the impacts. The methodology [7] used for the CS validation is composed of three different parts: the load evaluation, the CS strength evaluation and the final strength assessment comparing both the loads and the containment strength (see Figure 11). This approach is commonly accepted by Classifications Societies and also developed in their respective rules (ABS, 2009), (BV, 2011), (DNV GL, 2006) and (LR, 2009).

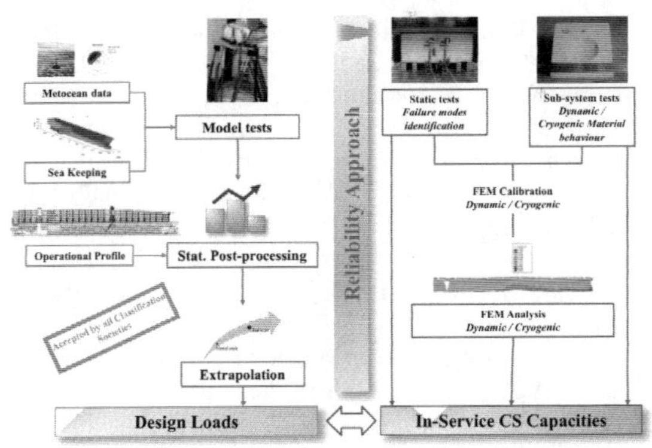

Figure 11: GTT Sloshing analysis process

The load evaluation is based on laboratory tests at small scale. For the time being it is impossible to make an accurate conclusion on CS reinforcement with numerical simulations. Indeed, the very localized effects occurring in the sloshing phenomenon induce a high variability of the loads and cannot be captured by current numerical models. Sloshing impacts are continuously recorded during tests thanks to pressure sensors fitted on the model tank. An event is measured when the pressure exceeds a given threshold. The maximum pressure of each event is analysed through statistical approaches. A long-term approach is considered for the sloshing study to determine the maximum expected sloshing loads on the CS wherever in the tanks. This approach integrates the sloshing response of all the sea states the ship will likely face during her life. The long-term pressure is extrapolated from the model scale to the full scale thanks to an empirical scaling law based on return of operating experience at sea.

The CS is submitted to both thermal (cryogenic conditions) and mechanical (dynamic loads) effects. In-service CS strengths are determined through experimental and numerical approaches. In laboratories, compressive tests are carried out on CS boxes or panels to evaluate their static strength. Sub-systems tests are performed to take into account thermal conditions and dynamic loads that will be used to calibrate the CS Finite Element Models (FEM). Indeed, thermal and dynamic effects are numerically calculated thanks to FEM. The combination of these two results defines the final in-service CS strength.

Loads and strengths are compared through a reliability method [7]. The results are presented in terms of annual probabilities of failure and depend on the nature of the considered structure.

For Mark III CS strength assessment, the main failure mode is the residual crushing of the foam. Whereas for NO96, the main failure modes are the failure of the primary boxes top covers and the buckling of the bulkheads' secondary boxes. The level of risk is obtained when the CS annual probability of failure is multiplied by the consequence associated with failure of the insulation system. For each failure mode, this risk has to be below an acceptance criterion defined as a frequency per year per ship. These criteria have been defined in agreement with ABS, BV, DNV GL and LR classification societies.

This assessment is performed for the two configurations of tanks – i.e. one-row and two-row – of a typical 16,000 TEU container ship in order to compare sloshing activity in both configurations and define the appropriate level of CS strength installed in the LNG fuel tank to withstand sloshing loads without filling limitation while the ship is sailing in North Atlantic.

3.3 SEA KEEPING ANALYSIS AND MAIN ASSUMPTIONS

Motions of the container vessel are necessary to issue time histories used for model tests. They are simulated with DIODORE software, using GTT standard methodology [8].

When ship motions become significant, waves are generated inside the tanks and hit the walls. The frequency and the magnitude of the impacts increase with the coincidence between the liquid resonant period and the natural period of the ship in relation with the sea state and heading. Sloshing is hence a resonance phenomenon depending on several parameters such as the environmental conditions, the ship and tank designs and the ship operational profile. These parameters are described in the next paragraphs.

Even if the typical journey for this type of container ship is Europe/Asia, we have conservatively assumed that the container vessel will spent 100% of its time sailing in North Atlantic Ocean with the worst filling level of LNG in the LNG fuel tanks. Indeed, North Atlantic conditions are identified as the most severe sea states which will conservatively cover all the worldwide navigation routes sailed by container ships. The selected wave spectrum is a Pierson-Moskowitz as recommended by IACS for North Atlantic.

Even if a container ship is more likely to operate at slow steaming to decrease its fuel consumption, the reference service speed of 23 knots is conservatively taken into account in the study up to 9 m of significant wave height. For higher wave heights, speed is reduced to account for drag increase in heavy seas.

3.4 EXPERIMENTAL SET-UPS

Two model tests campaigns have been carried out to assess sloshing in the tank(s) of a 16,000 TEU container ship, see [9]. The first one deals with a tank from a two-row arrangement. The second one concerns a one-row tank. The tests consist in moving a model tank at scale 1:40 with water and a heavy gas mixture composed of sulphur hexafluoride and nitrogen at ambient conditions. This gas composition is chosen in order to keep the same density ratio between liquid and gas as in a full scale LNG tank [10].

More than 200 pressure sensors are fitted in the model tank at locations where severe impacts are the most likely to occur. Pressures are measured for a given set of conditions – i.e. filling level, heading, wave period and significant wave height. The tank is moved thanks to a six-degree-of-freedom test rig named hexapod located in GTT laboratory (see Figure 12).

Figure 12: Two (among four) GTT six-degree-of-freedom test rigs

3.5 RESULTS OF MODEL TESTS

A different "worst filling level" has been identified in both model tests because of the different tank shapes. The critical filling level is around the height of the smaller lower chamfer – i.e. around 10%H for the two-row tank and 30%H for the one-row tank (see Figure 13), where H is the tank height at primary barrier level.

These low filling levels lead to progressive waves propagating from one vertical wall to the other and impacting on lower chamfers and longitudinal walls. The sloshing activity in the tanks is quantified by the events rates defined as the number of impacts per hour and the corresponding pressure magnitudes. Figure 14 shows the normalized events rates envelope measured in model tests campaigns for each heading on the one-row and two-row model tanks for their respective worst filling levels. Maximum events rates are obtained at beam seas (270°) for both tank configurations. Events rates are twice as high in the one-row tank than in the two-row tank. Hence, considering equivalent statistical laws, CS should be more reinforced in the one-row tank.

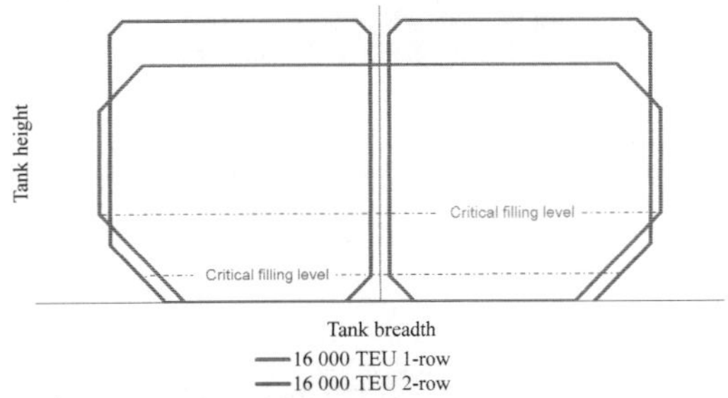

Figure 13: Critical filling levels

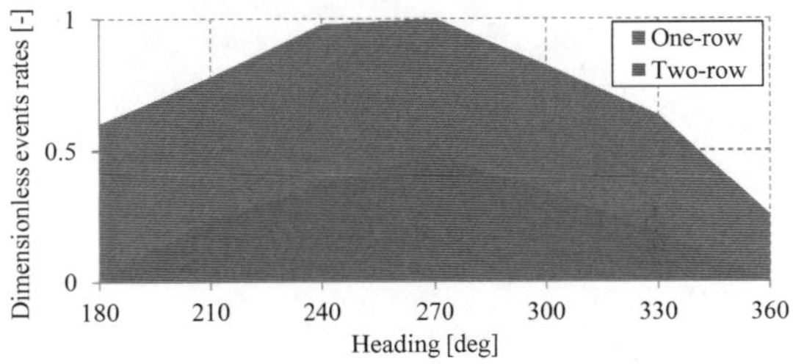

Figure 14: Dimensionless events rates comparison between one-row and two-row tanks

The fact that, at their respective "worst filling level", the events rates and the maximum pressures are generally higher in the one-row tank is due to the liquid transverse natural period which is closer to the roll natural period when the tank is larger. On the contrary, within a tank in a two-row arrangement, the resonance period between the liquid and the ship response is outside the critical filling range.

The sloshing assessments of both model test campaigns lead to the risk matrices presented in Figure 15 for Mark III and NO96 CS.

Pressures and events rates being higher in the one-row tank, a more reinforced CS is suitable compared to a two-row tank. Mark III with foam density of 130 kg/m^3 is appropriate for the two-row tank whereas 210 kg/m^3 foam is necessary in the one-row tank. As far as NO96 CS is concerned, Modified Ultra reinforced boxes are sufficient to withstand sloshing loads in the two-row tank. However a one-row tank should have to be fitted with NO96 Max boxes which are currently under development.

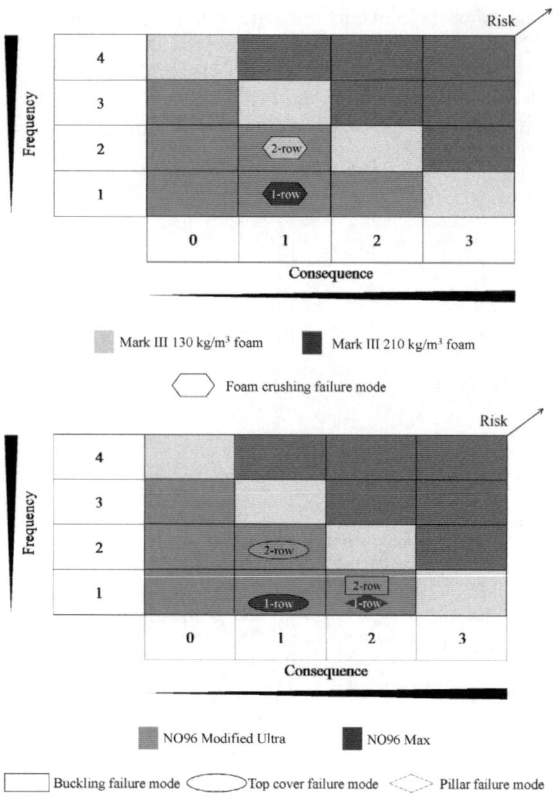

Figure 15: Risk matrices for NO96 and Mark III CS

The ability to operate the GTT membrane LNG fuel tank without filling limitation for a container ship in North Atlantic conditions is thus demonstrated through this assessment.

4. PRACTICAL SOLUTIONS FOR PROJECT EXECUTION

4.1 PROJECT ORGANIZATION

The experience of GTT is to provide support to a limited number of shipyards which are building large scale LNG carriers. When it comes to LNG fuel, the range of potentially involved shipyards is much wider with yards that have no experience at all at managing cryogenic systems.

In order to solve this matter, GTT proposes to work with outfitters to bring together the LNG expertise and assist the shipyards to design, procure and install the LNG fuel tank and related fuel gas handling system on board the different types of vessels. GTT already established industrial partnerships with Endel and Gabadi in Europe and AG&P in Asia. Others in China are under consideration.

4.2 SOLUTIONS FOR LNG FUEL TANK INSTALLATION AT NEWBUILDING OR RETROFIT

Several scenarios are possible to install LNG tanks for a new built as well as for a retrofit vessel. The membrane tank and fuel gas handling system can be installed with the conventional method, i.e. with installation of the CS when the inner hull is ready.

Another solution exists in order to minimize the newbuilding or retrofit lead time schedule. The LNG tank is fully manufactured separately as a tank block structure which has to be the exact structure of the ship. This tank block will include also an integrated room for fuel gas handling system. In case of retrofit, the inner structure of the cargo hold where the LNG tank will be implemented is cut-away with part of the double hull and double bottom structure. The edges of the remaining structure are prepared to accept the LNG tank. It is lifted with a large capacity floating crane and lowered down inside the space to be secured in place before being fully welded afloat (see Figure 17). This integration operation requires accurate positioning. The immobilization duration is estimated around four weeks.

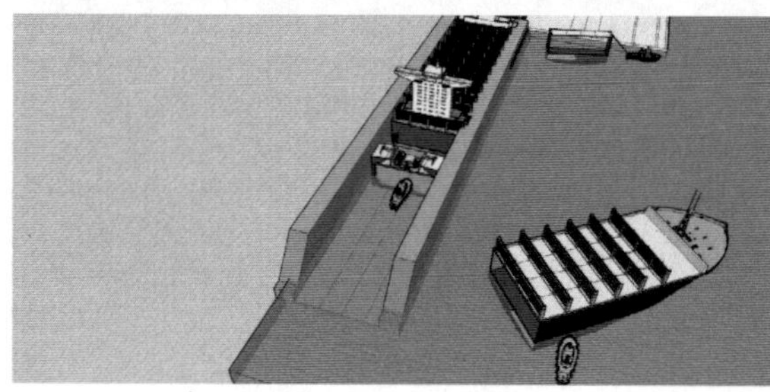

Figure 16: Installation of the new LNG membrane tank section

A third method to add an LNG membrane tank for a retrofit project consists in replacing the complete ship section where the LNG tanks will be implemented by a new one. This section is fabricated beforehand incorporating the LNG membrane tank with all the associated equipment. The vessel hull is cut in three parts (aft section, section to be removed and fore section) in a dry dock. They are then floated to remove the useless section thanks to tugboats. The new-built section is floated and moved inside the dry dock with the fore part and the aft part for the final assembly (see Figure 16).

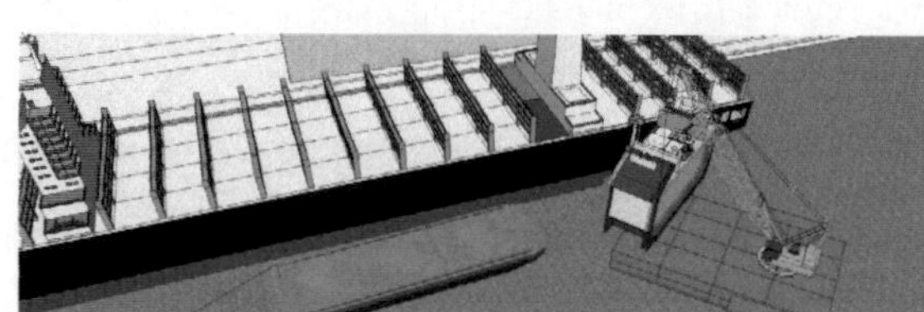

Figure 17: Integration of tank block structure

All these solutions have been studied with different partners such as ship designers, shipyards, ship-owners and classification societies for different sizes of container ships and other vessel types including cruise ships, ferries, ro-ro, VLOC, PCTC …

5. CONCLUSIONS

The requirements to reduce the NOx, SOx and particles emissions will lead the ship-owners to consider alternative fuels such as LNG. The consideration to change from HFO to LNG fuel tank on a typical large container ship has been evaluated with a GTT membrane tank technology.

It is demonstrated that even with an increased risk of collision compared to large LNG carriers, the integrity of the ship remain intact thanks to the high flexibility of the membrane tank under high deformation.

In addition, it is fully possible to install the adequate containment system configuration to operate in worldwide navigation without any filling restriction.

Container ships and other types of LNG fuelled vessels can now be built with membrane LNG fuel tank thanks to the support of GTT and its qualified industrial partners.

6. REFERENCES

1. P. CHARBONNIER, S. EHLERS, I. DARIE, 'Membrane containment system behaviour in case of collision', *LNG 18*, 2016.

2. IMO, 'Adoption of the international code of safety for ship using gases or other low-flashpoint fuels (IGF code)', *MSC.391(95)*, 2015.

3. 'SOLAS Consolidated Edition, Chapter II', *SOLAS*, 2014.

4. T. GLASER, 'Collision risk analysis for large container ships equipped with LNG fuel tanks – Master Thesis', *TUHH*, 2015.

5. P. JEAN, H. PETIT, 'Methane by Sea', 1998.

6. S. EHLERS, P. CHARBONNIER, I. DARIE, 'Experimental and numerical analysis of a membrane cargo containment system for liquefied natural gas', *ICSOS*, 2016

7. E. GERVAISE, P.E. DE SEZE, S. MAILLARD, 'Reliability-based methodology for sloshing assessment of membrane LNG vessels', *ISOPE*, 2009.

8. A. MARES, 'Refinement in sea-keeping analysis for sloshing investigation', *ISOPE*, 2009.

9. M.O. QUEMERE & al, 'LNG as a fuel on container vessels with GTT membrane systems', *ISOPE*, 2016.

10. S. MAILLARD, L. BROSSET, 'Influence of density ratio between liquid and gas on sloshing model test results', *ISOPE*, 2009.

7. AUTHORS BIOGRAPHY

Thierry Clément holds the current position of Business Development Manager at GTT. He is responsible for business development in Scandinavia area and some key accounts in other areas. He previously occupied various positions including R&D Engineer in Hydrodynamics, Head of Liquid Motion Department and Project Manager in charge of the relations with Shipbuilding Yards (China, Korea, Japan). He has been involved in liquid motion analysis and the determination of sloshing loads on membrane-type insulation systems. He principally worked on the study of LNG carrier projects as well as offshore projects.

Pierre Charbonnier holds the current position of Technical Project Leader at GTT. He is responsible for internal projects as the collaboration with the Technical University of Hamburg and DNV GL about the collision with a container ship. His previous experience includes the arctic validation of the containment system for the first LNG carrier - ice breaker.

Marie-Odette Quéméré holds the current position of Hydrodynamics Engineer at GTT. She is responsible for liquid motion studies by carrying out sloshing analyses and model test campaigns.

SPECTRAL FATIGUE ANALYSIS OF LIQUEFIED NATURAL GAS CARRIER STRUCTURAL DETAIL

N Vladimir and **I Senjanovic,** University of Zagreb, Croatia
S Malenica and **C Ouled Housseine,** Bureau Veritas, France
BK Choi and **HI Im,** Hyundai Heavy Industries, South Korea

SUMMARY

The spectral fatigue analysis of a structural detail of Liquefied Natural Gas (LNG) carrier is presented. The analysis was performed by means of general hydro-structure tool HOMER (BV), where 3D FEM model for the structure and 3D potential flow code for fluid modelling, respectively, is applied. Mode superposition method is used to calculate ship hydroelastic response in waves. Numerical procedure for the fatigue assessment based on the so called top-down procedure is described in details and applied to determine stress concentrations in fine mesh FE model of a selected structural detail. Based on the calculated stress RAOs and taking into account proper S/N curve, the fatigue life of certain structural detail is obtained for a selected operational profile.

1. INTRODUCTION

Natural gas is considered as a cleaner fuel compared to MDO or HFO. In line with increasing environmental care, there is a rising demand for natural gas worldwide and international natural gas markets are continuously growing. One of the most important parts of LNG transportation system are LNG ships, which can be found in service from 1959. A typical LNG carrier is double-hull vessel with four to six tanks located along the centreline, and there are nowadays several containment systems in use, that can be classified into self-supporting ones (also referred to as Moss type) and membrane type ones. Both alternatives are designed, constructed and equipped with sophisticated systems for carrying LNG over long distances at storage temperatures around -162 °C [1]. Recently, the membrane tank system has been adopted widely due maximizing ship load capacity. More details on LNGC cargo containment systems can be found in [1].

IHS Fairplay (IHSF) database [2] includes data on all ships operating worldwide, and here the DWT of ships delivered from 1999 to 2015 (inclusive) is presented in Figure 1, which shows us that largest LNG ships have been built about 10 years ago (several ships with DWT smaller than 60000 t are omitted from the representation). However, it is more interesting to look at the number of delivered ships and number of orders by the end of 2019 (as stated in August 2015), Figure 2. Although there is some slight trend to build larger units, it seems that market of LNG ships is rather unstable, which is a consequence of economic crises.

There are different issues associated with the design (cargo containment system, hydrodynamic aspects, structural aspects, propulsion issues...) and operation (LNG transfer systems, partial filling issues, problem of boil-off…) of LNG ships making them rather complex objects, and this paper is oriented to the structural one, i.e. how to assess fatigue life of a ship structural detail within so-called direct calculation approach.

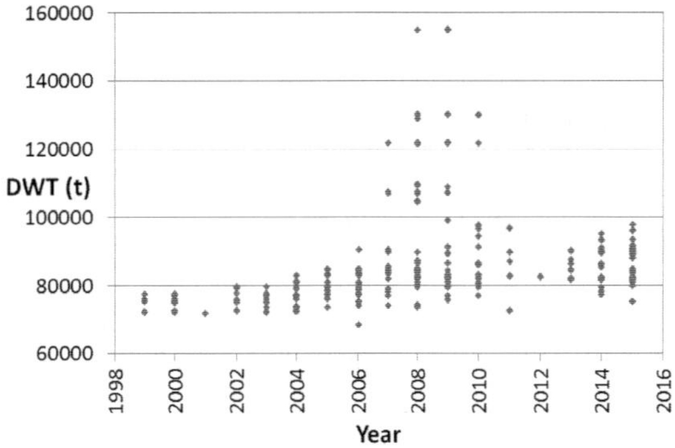

Figure 1: DWT of LNG ships delivered from 1999 to 2015

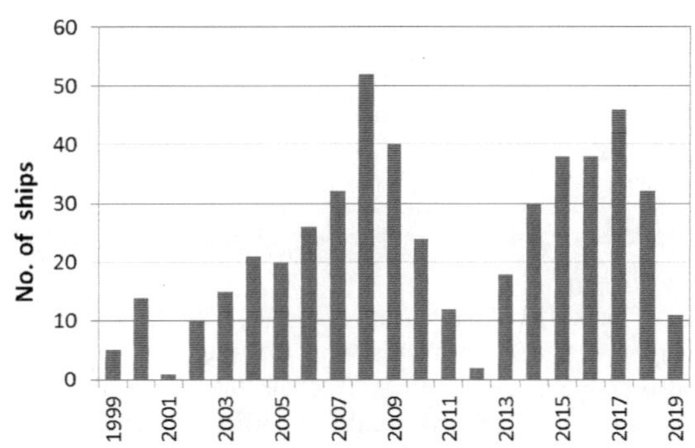

Figure 2: LNG ship fleet and order book

The applied procedure is elaborated in details, as well as the used general hydro-structure tool HOMER [3,4]. More information on the mentioned software is given within the paper, but it is to be noted that it can be applied to any kind of ships and offshore structures in the analysis of both quasi-static and dynamic structural responses, caused by linear, weakly nonlinear or impulsive nonlinear hydrodynamic loading, respectively. Although hydroelastic effects are not expected to be of primary importance in this case (due to relatively high hull girder rigidity and low operational speed), computation of stress RAOs needed for fatigue damage computation is done by hydroelastic model (linear springing included) taking into account also the effect of internal liquid in the cargo tanks, just to illustrate applicability of the software within the complete procedure.

2. DESCRIPTION OF THE ASSESSMENT PROCEDURE

Linear hydroelastic analysis performed here is based on the mode superposition method [5]. Within the modal approach, total displacement of a ship is expressed through a series of modal displacements:

$$\boldsymbol{H}(x,t)=\sum_{i=1}^{N}\xi_i(t)\boldsymbol{h}^i(x), \tag{1}$$

where $\boldsymbol{H}(x,t)$ represents total displacement of one point, $\boldsymbol{h}^i(x)$ is modal displacement (mode shape), $\xi_i(t)$ is modal amplitude, and N represents the total number of modes [4]. Generally, the procedure is very similar to rigid body analysis described in [6] except that the number of degrees of freedom is extended from 6 to 6 plus a certain number of elastic modes. The used modal approach implies the definition of supplementary radiation potentials with the following body boundary condition:

$$\frac{\partial \varphi_{Rj}}{\partial n}=\boldsymbol{h}^j\boldsymbol{n}, \tag{2}$$

where $\boldsymbol{n}$ is unit normal vector. After solving the different boundary value problems for the potentials, the corresponding forces are calculated and the equation of motion is written

$$\left\{-\omega^2(\mathbf{m}+\mathbf{A})-i\omega(\mathbf{B}+\mathbf{b})+(\mathbf{k}+\mathbf{C})\right\}\boldsymbol{\xi}=\mathbf{F}^{DI}, \tag{3}$$

where $\mathbf{m}$ is the modal structural mass, $\mathbf{b}$ is the structural damping, $\mathbf{k}$ is the structural stiffness, $\mathbf{A}$ is the hydrodynamic added mass, $\mathbf{B}$ is the hydrodynamic damping, $\mathbf{C}$ is the hydrostatic restoring stiffness, while $\mathbf{F}^{DI}$ is the modal hydrodynamic excitation vector.
Once the modal amplitudes have been calculated the total stresses can be obtained, at least theoretically, by summing the individual modal contributions and one can formally write, [6]:

$$\Sigma(x,\omega)=\sum_{i=1}^{N}\xi_i(\omega)\sigma^i(x), \tag{4}$$

where $\Sigma(x,\omega)$ is the total stress and $\sigma^i(x)$ is the spatial distribution of modal stresses.
In order to practically take into account hydroelastic effects on the structural response, dynamic computational scheme is applied, starting with modal analysis in dry condition, Figure 3, [7]. Once the dry modes are obtained, the modal displacements are transferred from the structural model to the hydrodynamic one, and corresponding hydrodynamic problem is formulated. After that, fully coupled dynamic equation is solved, giving the modal amplitudes.

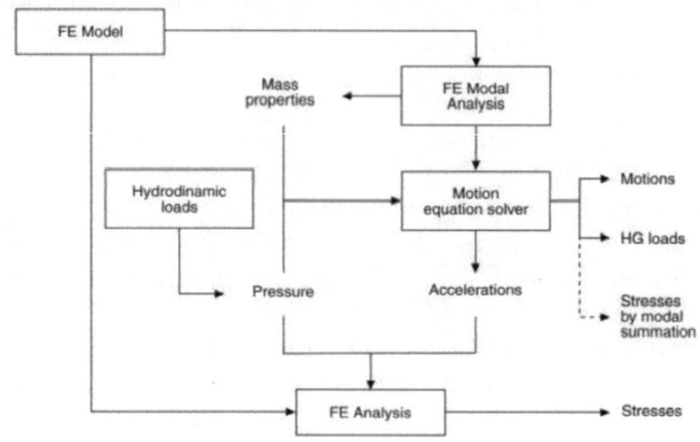

Figure 3: Dynamic analysis computational scheme [7]

In order to cover all types of hydro-structural interactions inherent ships and offshore structures described in [4], the numerical software HOMER is developed in Bureau Veritas Research Department for the direct transfer of the seakeeping loads from the general seakeeping code to a structural FE model, Figure 4, [3,4]. HOMER modules presented in Figure 4 are intended to be used as follows: HMFEM – to compute mass and inertia properties of FE model. Run modal analysis, HMSWB – to analyse still water load case and perform balancing, HMHST – for running hydrodynamic pressures computations using the seakeeping code, HMMCN - solves mechanical problem, HMFEA – to run FE analysis on load cases, HMRAO - to create RAOs and HMTIME – to perform time-domain computations.

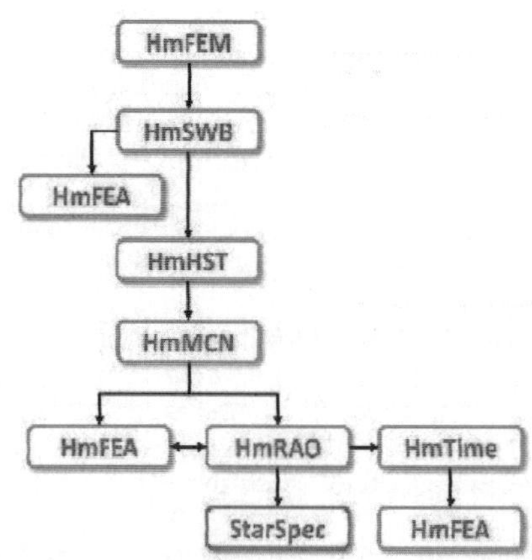

Figure 4: Flowchart of HOMER software application [3]

Three main ideas introduced through HOMER software to obtain the perfect equilibrium of the structural model are the following, [4]:

1. Recalculation of the pressure at the structural points (instead of interpolation).

2. Separate transfer of the different pressure components, and calculation of the different hydrodynamic coefficients by integration over the structural FE mesh.
3. Solution of the motion equation using the above calculated hydrodynamic coefficients and inertia properties of the FE model. This point ensures the perfect equilibrium of the FE load case because of calculation of all the coefficients of the motion the FEM model.

Within the investigation presented in this paper, HOMER is used with Hydrostar [8] as the hydrodynamic solver, and NASTRAN [9] as the structural solver.

Fatigue assessment of LNG ship structural detail is performed according to the flowchart presented in Figure 5.

For the fatigue life/damage calculation, very local stress concentrations in some particular structural details are needed, and generally they can be calculated by refining the global coarse mesh or using the so called top-down approach. The former approach seems to be impractical leading to excessive number of finite elements, and therefore here, the latter one is used, which implies solving the global coarse mesh FEM problem at first, and applying the coarse mesh displacements at the boundaries of the local fine mesh later [10]. In this way the fine mesh FEM calculations are performed in a next step with the load cases defined by the prescribed displacements from the coarse mesh and by the local pressures and inertia of the fine mesh. The above procedure should be performed for each operating condition, defined by loading, wave frequency and heading, and for both real and imaginary part of the loading, resulting in the RAOs of the stresses in each particular structural detail.

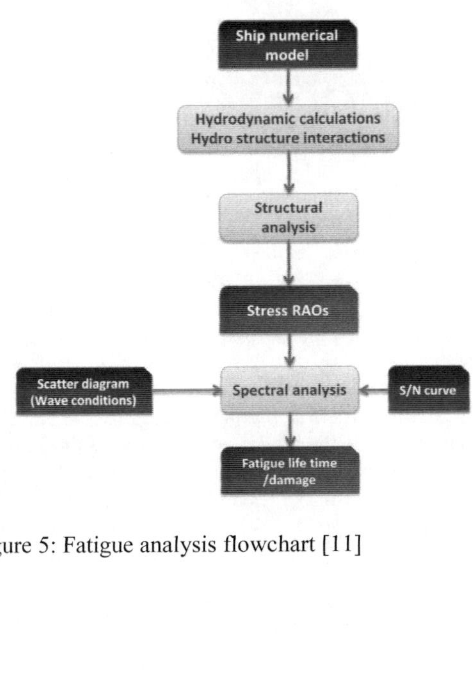

Figure 5: Fatigue analysis flowchart [11]

A special care is given to the separation of the quasi-static and dynamic parts of the response to ensure a proper convergence of the results. The quasi-static part of the response is calculated using the so called quasi-static method as described in [7], and dynamic part of the response is calculated by summing up the dynamic contribution of each mode.

After obtaining transfer functions of stresses, the spectral analysis is performed and based on the selected S/N curve and wave scatter diagram, fatigue life/damage is calculated [11].

Finally, it is to be noted that the procedure for the fatigue assessment is the same irrespective on calculation of stress RAOs in quasi-static or hydroelastic manner.

3. SHIP DATA, NUMERICAL MODELS

Application of HOMER software to fatigue assessment of ship structural detail is illustrated on 175 KM3 LNG vessel, with tank general arrangement shown in Figure 6. The main ship particulars are the following:

Length between perpendiculars:	$L_{pp} = 282.0$ m
Breadth:	$B = 46.0$ m
Depth:	$H = 26.0$ m
Draught:	$T = 11.6$ m
Displacement, full load:	$\Delta_f = 116973.0$ t
Displacement, ballast:	$\Delta_b = 92194.6$ t

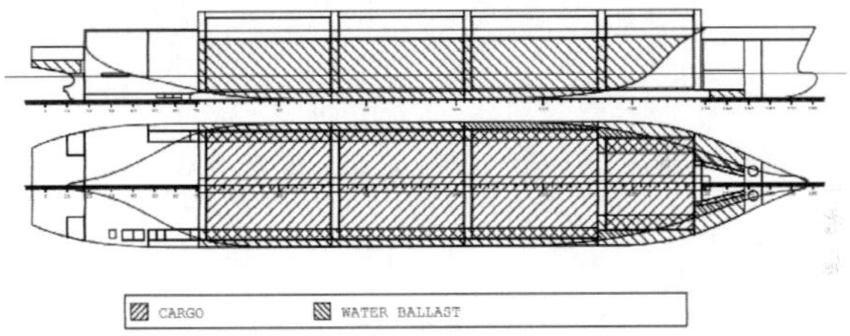

Figure 6: General arrangement of ship cargo tanks

Global FE model of the considered ship, having 427752 elements is used for the calculation, Figure 7. Actually, much coarser mesh can be used for this purpose, but one of the aims of this investigation was software testing at very demanding finite element models.

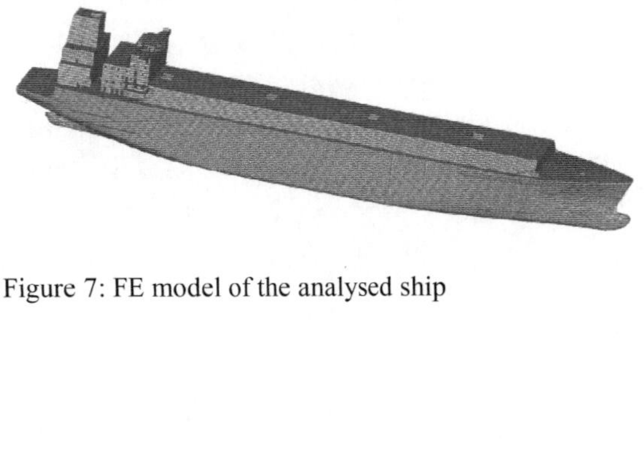

Figure 7: FE model of the analysed ship

Beside global FE model, fine mesh model of a structural detail is required, to be included in the calculation within the top-down scheme, Figure 8.

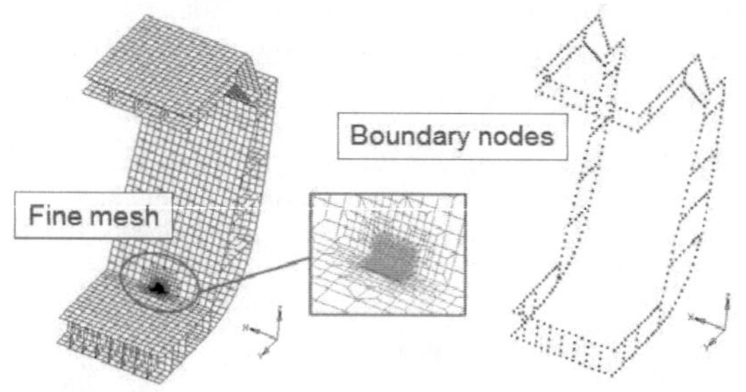

Figure 8: Fine mesh FE model with selected boundary nodes for top-down procedure

The fine mesh model is located in the fore tank, which is for the purpose of this calculation set at 70% of filling capacity, while other 3 tanks are full, Figure 9. It should be noted that this loading condition is not one of the standard ones from loading manual and the ship is not aimed to operate in this way, but it is used here in an academic way for the purpose of investigation.

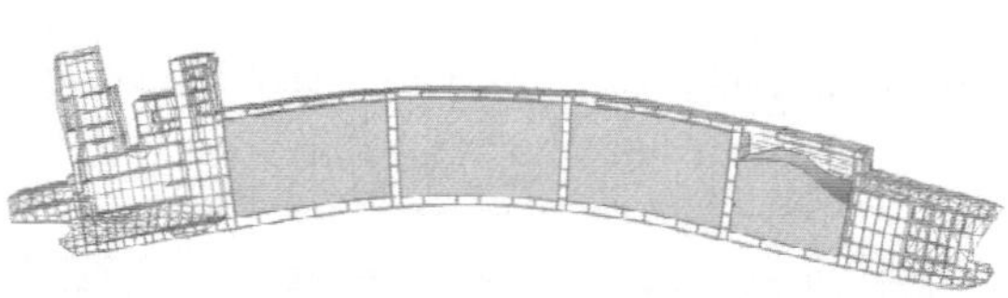

Figure 9: Illustration of the selected loading condition on deformed ship structural model

Beside both FE global and local models of a ship structure, applied procedure also requires generation of the so called integration and hydrodynamic mesh, respectively, Figures 10 and 11. The former is extracted directly from the structural model, and then the latter one, is generated automatically using the existing software routines.

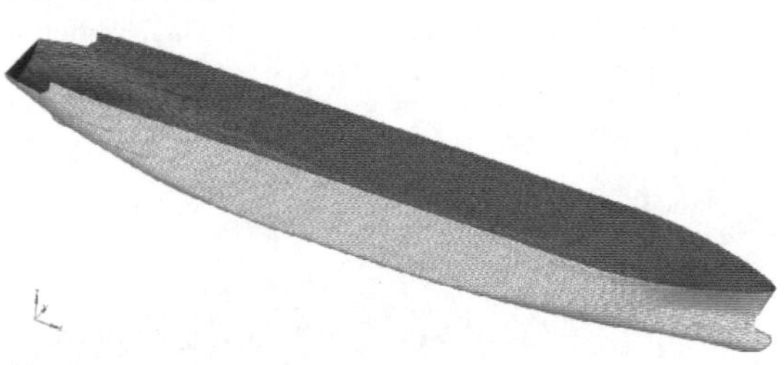

Figure 10: Integration mesh

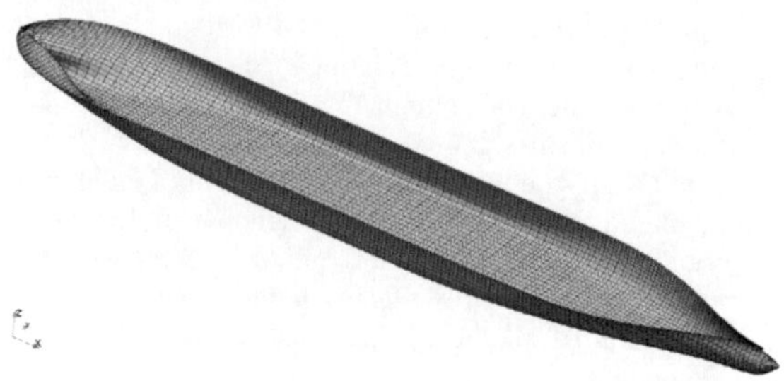

Figure 11: Hydro mesh

The same meshes are required for all cargo tanks containing liquid. Example of tank integration mesh that is also used as a hydro mesh is shown in Figure 12.

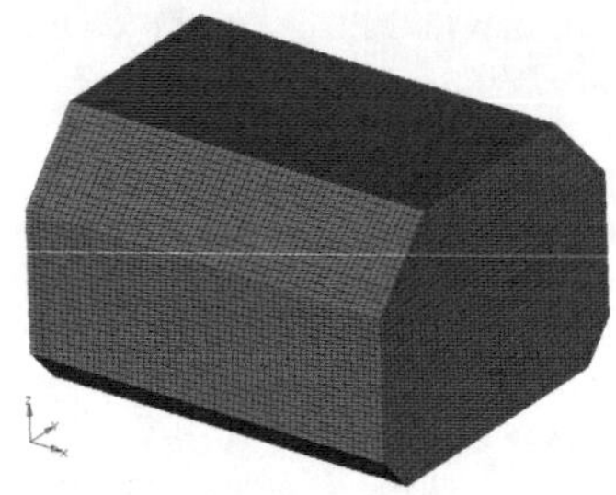

Figure 12: Tank integration and (hydro) mesh

Finally, for the analysed detail, both internal and external integration meshes are needed, Figure 13.

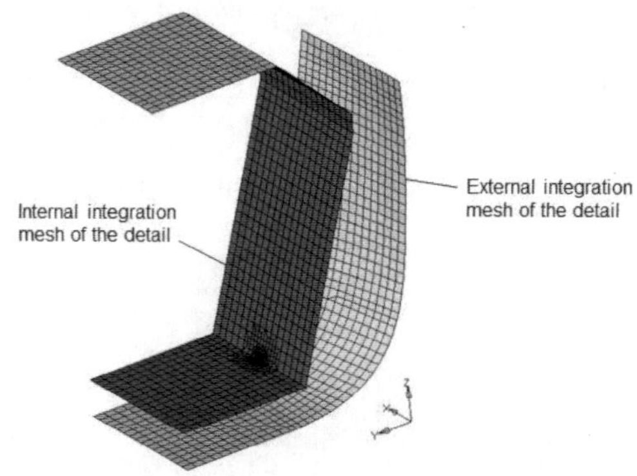

Figure 13: Detail integration meshes

4. RESULTS

Due to reason of simplicity, the analysis was performed for zero forward speed, while wave headings are considered uniformly distributed from 0° to 350° with step of 10.0°. The range of wave frequencies is set from 0.0 to 2.0 rad/s with a step equal to 0.05 rad/s. The first computation step is the modal analysis giving dry natural modes and frequencies, Figure 14, whereas 8 elastic modes are retained for hydroelastic computations.

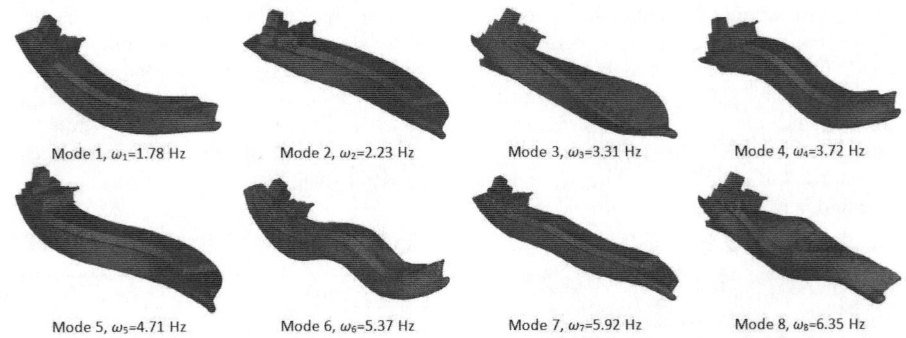

Figure 14: Dry natural modes and frequencies of analysed LNG ship

Prior to running hydrodynamic computations it is very useful to analyze still water case results, that serve as recommended checks of structural and hydrodynamic model consistency, their relative positions in global coordinate system, mass modelling within the structural model and basic calculation setup, respectively, etc. Just for the illustration, here the hydrostatic pressures on ship hull, tank and structural detail are given showing the values accurately reflecting given draught, Figures 15-17.

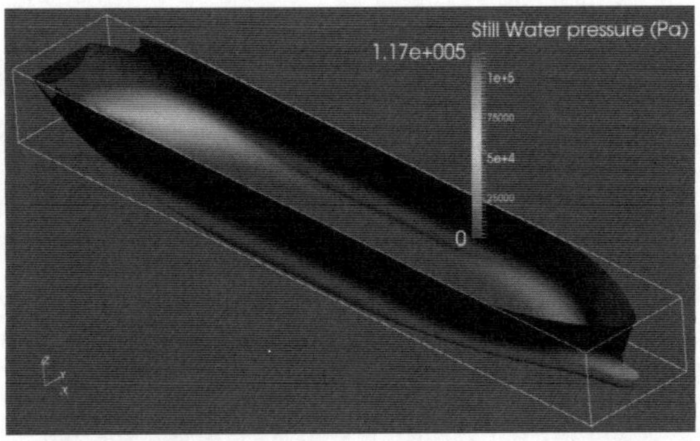

Figure 15: Hydrostatic pressures on hull

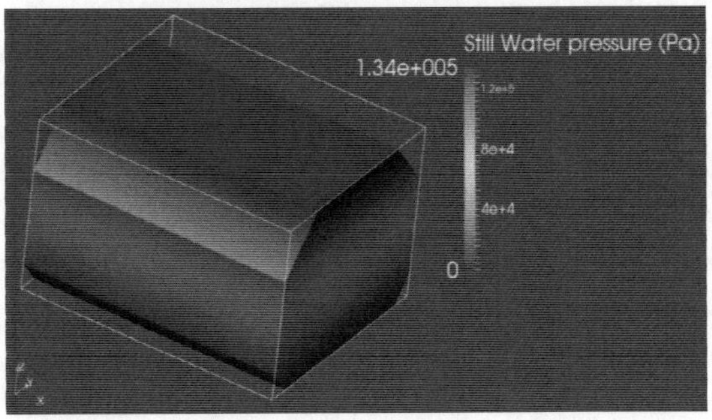

Figure 16: Example of hydrostatic pressures on tank

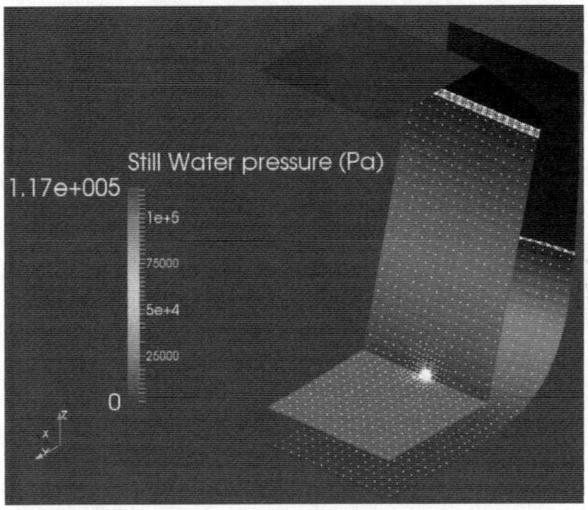

Figure 17: Hydrostatic pressures on detail

Figures 18 and 19 show still water deflections and stresses, where also reasonable numerical values are obtained.

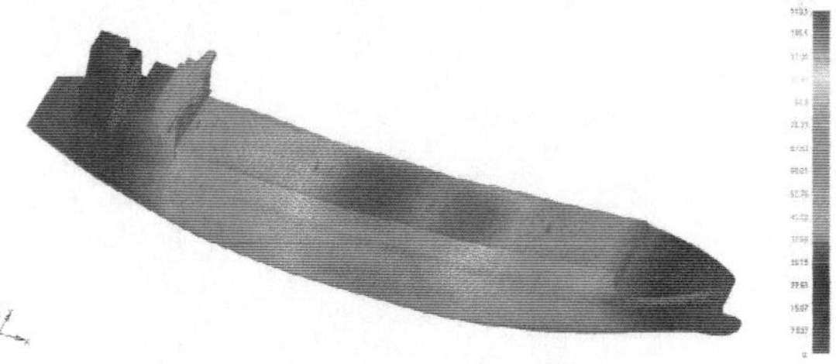

Figure 18: Still water deflections (mm)

Figure 19: Still water stresses (N/mm^2)

After performing hydrodynamic computations and solving equation of motion, the global hydroelastic response is obtained. The global results are imposed to detail FE model and FE analysis for each combination of ship speed, wave frequency and heading, series of stress RAOs is done. Example of stress pattern in a structural detail is shown in Figure 20, while typical stress RAO is presented in Figure 21.

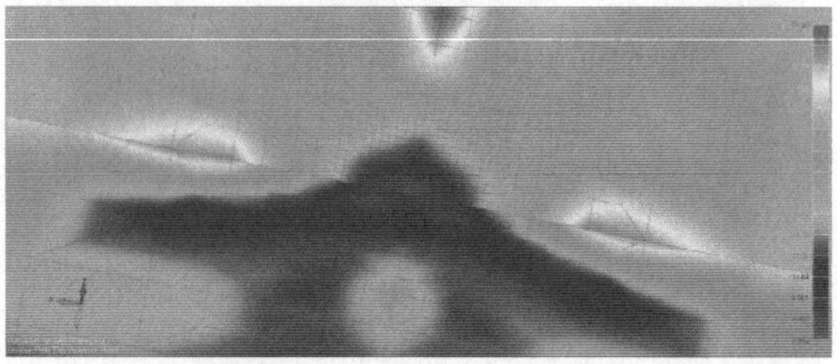

Figure 20: Example of Von Mises stresses (N/mm^2) in a ship structural detail

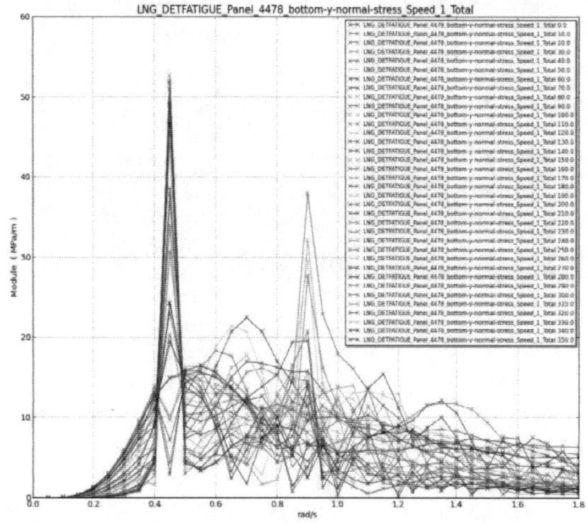

Figure 21: Typical stress RAO

Fatigue lives are computed for all finite elements in the very fine mesh model, and here several ones with the lowest fatigue lives are only identified, Figure 22.

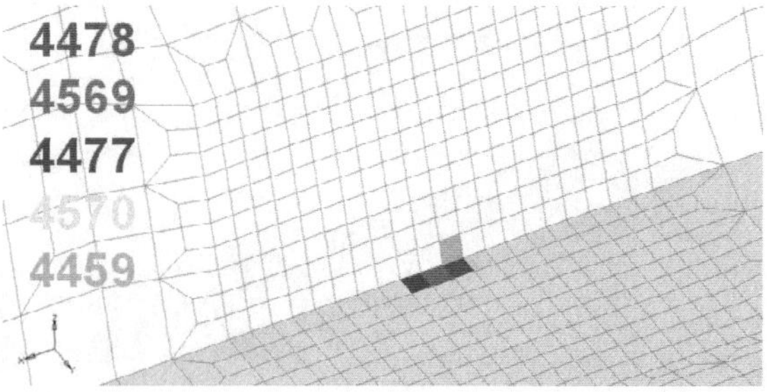

Figure 22: Identification of finite elements with lowest fatigue lives

Fatigue lives for the above finite elements are listed in Table 1, and indicate good fatigue performance of a ship. However, more detailed analysis with realistic ship loading conditions is required to offer final conclusions on her fatigue strength, since this calculation is intentionally performed in a rather academic manner.

Table 1: Fatigue lives of selected finite elements

Element ID	**Fatigue life (years)**
4478	32.5
4569	33.7
4477	59.0
4570	61.4
4459	81.5

5. CONCLUSIONS

Spectral fatigue assessment procedure by means of general hydro-structure tool 3D FEM structural model and 3D BEM hydro model, respectively, is described in details. All necessary calculation steps and required numerical models are discussed, as well as top-down procedure for the assessment of local stresses, which are further used for the fatigue damage computation. Since fully coupled hydroelastic model is used, ship elastic deformations and her motions in waves are simultaneously taken into account. Fine mesh model of a tank structural detail of LNG carrier, exposed to both internal and external fluid pressure, was used as an application case. The obtained results indicate that the analysed structural detail is well designed from the viewpoint of fatigue, which is in accordance with the fact that analysed ship is in service for several years with no reported damage for the considered detail. Also, the software was found to be user friendly and very robust for application to fatigue assessment of ships and offshore structures, in spite of the very large finite model analysed in this investigation. After these findings, the future work will be oriented to fatigue analysis of the considered ships with forward speed and a set of more realistic loading conditions.

6. ACKNOWLEDGEMENTS

This work was supported by a National Research Foundation of Korea (NRF) grant funded by the Korean Government (MSIP) through GCRC-SOP (Grant No. 2011-0030013) and by University of Zagreb, Croatia within the short-term research grant: "Safety, Energy Efficiency and Environmental Eligibility of Liquefied Natural Gas Carriers". The first author would also like to acknowledge support of Bureau Veritas for his 3-month visit to their Research Department in Paris. The obtained results and conclusions reflect personal opinions of the authors and do not represent official attitude of their institutions.

7. REFERENCES

1. KIM, H.S., CHUN, M.S., LEE, J.M., KIM, M.H., 'A comparative evaluation of fatigue and fracture characteristics of structural components of liquefied natural gas carrier insulation system', *Journal of Pressure Vessel Technology*, Vol. 135, pp. 1-9, 2013.

2. IHS, 'World Register of Ships – WROS manual', 2015.

3. SIRETA, F.X., DE LAUZON, J., SURMONT, F., 'HOMER User guide', Bureau Veritas, Paris, France, 2013.

4. MALENICA, S., DERBANNE, Q., SIRETA, F.X., BIGOT, F., TIPHINE, E., DE HAUTECLOCQUE, G., CHEN, X.B., 'HOMER – Integrated hydro-structure interactions tool for naval and offshore applications', *International Conference on Computers Application in Shipbuilding*, Busan, Korea, 2013.

5. BISHOP, R.E.D., PRICE, G., 'Hydroelasticity of Ships', Cambridge University Press, 1979.

6. MALENICA, S., SENJANOVIC, I., VLADIMIR, N., 'Hydro structural issues in the design of ultra large container ships', *Brodogradnja*, Vol. 64, No. 3, pp. 323-347, 2013.

7. BUREAU VERITAS, 'Whipping and Springing Assessment', Rule Note NR 583 DT R00 E., Paris, France, 2015.

8. BUREAU VERITAS, 'Hydrostar, User's manual', Paris, France, 2006.

9. MSC SOFTWARE, 'MD Nastran 2010 Dynamic analysis user's guide', Newport Beach, California, USA, 2010.

10. SIRETA, F.X., DERBANNE, Q., BIGOT, F., MALENICA, S., BAUDIN, E., 'Hydroelastic response of a ship structural detail to seakeeping loads using a top-down scheme', *31st International Conference on Ocean, Offshore and Arctic Engineering - OMAE*, ASME, Brazil, 2012.

11. IM, H.I., VLADIMIR, N., MALENICA, S., RYU, H.R., CHO, D.S., 'Fatigue analysis of HHI SkyBench^{TM} 19000 TEU ultra large container ship with springing effect included', *7th International Conference on Hydroelasticity in Marine Technology*, Split, Croatia, 2015.

8. AUTHORS BIOGRAPHY

Nikola Vladimir is Assistant Professor at Faculty of Mechanical Engineering and Naval Architecture, University of Zagreb (UNIZAG), Croatia. He obtained MSc degree in Naval Architecture in 2007 at UNIZAG, and PhD in the field of Hydroealsticity of ship structures in 2011 at the same University. His research interests include vibration of ship structures (hydroelastic – springing & whipping, engine & propeller induced, etc.), vibration phenomena in ship propulsion system (main engine, shafting, propeller…), fatigue, specialized software development and application as well as all other dynamic loading and response issues inherent to ships and offshore units. He published over 100 papers in journals and specialized conference proceedings. Member of Dynamic Response Committee of ISSC.

Ivo Senjanovic, Professor Emeritus of Naval Architecture at the University of Zagreb, Faculty of Mechanical Engineering and Naval Architecture, Zagreb, Croatia. Teaching several courses on strength and vibration of ships, offshore structures and submarines. Investigations fields are ship stability, launching, ship strength and vibration, shell theory, design of pressure vessels, non-linear dynamics, ship hydroelasticity, numerical methods, numerical simulations, etc. Books: "Theory of Plates and Shells", "Finite Element Method in Ship Structure Analysis", "Ship Vibrations". Published over 300 papers in scientific journals and conference proceedings. ISSC member acting in Technical Committees on ship strength, vibration, load and response during 10 three-year periods. Fellow of the Croatian Academy of Sciences and Arts.

Sime Malenica graduated in Naval Architecture at the Faculty of Mechanical Engineering and Naval Architecture, University of Zagreb in 1990. He obtained his PhD degree in 1994, at the University Paris VI, on the problem of nonlinear water wave diffraction. He has published more than 100 papers both in scientific journals and at the specialized conferences. Dr. Malenica is Deputy Director of Bureau Veritas Research Department, leading several national and international research projects. His main fields of research concern the hydro-structural interactions due to the sea waves (linear, nonlinear, frequency domain, time domain, ships and offshore structures…). Actual topics concern the hydroelasticity issues for ships (springing, whipping, slamming, sloshing, green water…). He is a member of editorial board of several respectable journals, Dynamic Response Committee member of ISSC and corresponding member of Croatian Academy of Sciences and Arts.

Charaf Ouled Housseine received MSc degree in Hydrodynamics & Ocean engineering in 2014 from Ecole Centrale de Nantes (ECN), France. Since then, he has been working in Bureau Veritas Marine & Offshore Division, France, as a hydro-structure research engineer. He has been in charge of developing simulation tools and performing advanced studies of ships and offshore unit hydroelastic response. His current research interests include mainly fluid-structure interaction under different approaches: rigid, elastic, linear, nonlinear, etc., as well as other hydrodynamic loads such as Morison forces and linear sloshing.

Byung-Ki Choi, PhD, holds a position of head researcher in the Structure Research Department, Hyundai Heavy Industries. He is responsible for structural assessments of ships and MODU structures. Member of Fatigue and Fracture Committee of ISSC.

Hong-Il Im currently holds a team leader position as a senior engineer in basic hull design department of Hyundai Heavy Industries. Specialized in evaluation of hull girder vibration including offshore, basic hull design of container carriers, tankers, offshore structures and special purpose vessels.

INFLUENCE OF LIQUID CARGO ON FLNG MOTION

A Nakashima, JGC Corporation & Yokohama National University, Japan
M Arai, Yokohama National University, Japan

SUMMARY

This paper introduces a spring-mass type mechanical model to represent the liquid cargo behaviour and the coupling effects on FLNG motion. The mechanical model of sloshing was originally developed in the mid-1960s and used to represent the coupling effects on the spacecraft motion for predicting and controlling the spacecraft behaviour and stability. Authors believe that the mechanical model is a useful tool for better understanding the interaction of liquid cargo and FLNG motion, and it is also a simple solution for the early design stage to determine the principal FLNG configurations.

First the concept and associated formula of the mechanical model are explained in this paper. Then the parameters of the mechanical model for a rectangular tank are obtained analytically and parameter determination by numerical or experimental method for arbitrary shaped tank is explained. The equation of floating body motion and the mechanical model are coupled, and finally numerical demonstrations are performed and the results are verified by the literature.

NOMENCLATURE

a	Dimension along x axis of rectangular tank (m)
b	Dimension along y axis of rectangular tank (m)
g	Gravity constant (9.80665 m s^{-2})
H_0	Fixed mass height (m)
H_n	Slosh mass height (m)
K_n	n-th spring constant (N m^{-1})
h	Liquid height (m)
I_0	Fixed mass inertia (kg m^2)
I_R	Mass inertia of rigid cargo (kg m^2)
I_d	Mass inertia of free rotating liquid (kg m^2)
m_0	Fixed mass (kg)
m_n	Slosh mass (kg)
m_T	Total liquid mass (kg)
M	Mass of floating body including liquid cargo (kg)
M_B	Mass of floating body excluding liquid cargo (kg)
p	Fluid Pressure (N m^{-2})
γ_n	n-th fraction of critical damping (-)
X_0	Linear displacement along x axis (m)
x_n	Deflection of the slosh mass (m)
(x_G, y_G, z_G)	COG of floating body including liquid cargo (m)
(x_{BG}, y_{BG}, z_{BG})	COG of floating body excluding liquid cargo (m)
(x_{TG}, y_{TG}, z_{TG})	COG of liquid cargo (m)
α_0	Angular rotation about an axis through the COG of liquid cargo (rad)
ρ	Density of surrounding water (kg m^{-3})
ρ_c	Density of cargo liquid (kg m^{-3})
Φ	Velocity potential (m^2 s^{-1})
ω	Angular frequency (rad s^{-1})
ω_n	n-th natural angular frequency (rad s^{-1})

1. INTRODUCTION

During the past decade, the demand of liquefied natural gas (LNG) has increased sharply and is expected to further increase in near future, as cleaner energy comparing to the other fossil fuel sources (e.g. oil and coal) and more stable supply energy contrasted with the renewable energy sources (e.g. wind, solar, etc.). Governments and major oil and gas companies have a lot of interest in the LNG development. This has resulted in the technical innovation in LNG supply chain, such as the capacity increase of LNG tank system, the development of floating LNG production, storage and offloading unit (FLNG), floating storage and regasification unit (FSRU). FLNG has attracted high expectations for the development of offshore stranded gas fields or monetizing associated gas from oil fields.

One of the important problems to be solved for realization of FLNG is the motion prediction technique for floating body/bodies. This technique is essential to estimate the reliability and availability of FLNG. Research and development of the motion prediction technique has been performed by a number of researchers and research groups for a long period of time. In combination with the increased computing power, the motion prediction techniques have achieved a sufficient level for practical application. Even though there are still technical issues in improving the accuracy of motion prediction, such as the incorporating method of viscous effect, narrow gap resonance, liquid cargo effects, wind and current load estimation.

Authors focus on the liquid cargo effects on the FLNG motion in this study. Several numerical methods for dynamic coupling between liquid motion in tanks and rigid ship motion have been studied and established [1 to 10]. The liquid cargo behaviour is represented by the fluid dynamic methods such as boundary element method, finite difference method and particle method, and it is coupled with the floating body motion. These methods can simulate the dynamic behaviour of liquid and ship with sufficient accuracy, and were confirmed by comparing with model test results.

However, those advanced numerical analysis is too complex and black box for non-researchers, i.e. industrial users, and it requires a large computational demand. Since there is no clear view of liquid cargo behaviour and its coupling effects to the floating body motion, industrial users must verify the analytical result by model test.

Those are the barriers for practical application. The model test and the advanced numerical motion analysis including sloshing effect can be carried out in late design stage.

On the other hand, important FLNG configuration will be determined based on a simple analysis at the early design stage. The advanced numerical analysis calibrated by model test will be carried out later and it will verify the soundness of the design, requests the local modification, reinforcement, or suggests limitation or procedure of operation. If significant design change is requested by the model test or advanced numerical analysis, it is not practical to incorporate such change due to the project schedule and budget. Therefore, the design in early stage is definitely important. Underestimation will cause the catastrophe of the project, and overestimation will result in the loss at the job competition.

At this moment, FLNG motion affected by liquid sloshing can only be considered qualitatively for determination of important FLNG configuration. If quantitative sloshing assessment is carried out in the early design stage, there is more possibility to take cost effective tank configuration, such as 1 row membrane tank.

In the late design stage, offloading down-time simulation [11] will be carried out. Even the state-of-the-art simulation does not incorporate the liquid sloshing. One reason may be that the liquid sloshing has not been paid attention in FLNG motion due to the adoption of the 2 row tank configuration or SPB tanks. Another reason is the large computational demand.

According to that problem awareness, authors intend to develop another simple and less computationally demanding motion analysis method that can consider liquid sloshing.

2. MECHANICAL MODEL OF LIQUID CARGO

2.1 GENERAL

The main dynamical effect of lateral sloshing is a horizontal oscillation of the liquid centre of mass relative to the tank. This effect can be well represented by an equivalent mechanical model as shown in Figure 1. The mechanical model of sloshing was originally developed in the mid-1960s and used to represent the coupling effects with the spacecraft motion for predicting and controlling the spacecraft behaviour and stability [12, 13].

The pendulum model was adopted in the spacecraft application since the gravity acceleration is treated explicitly as a variable parameter as shown in Figure 2. However, the spring-mass model is a better choice for the marine and offshore industry application because the change in gravity acceleration is not necessary to be considered.

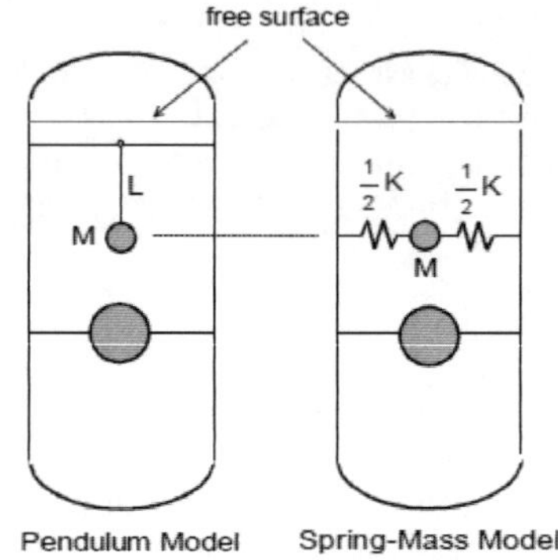

Figure 1: Mechanical Model of Liquid Cargo

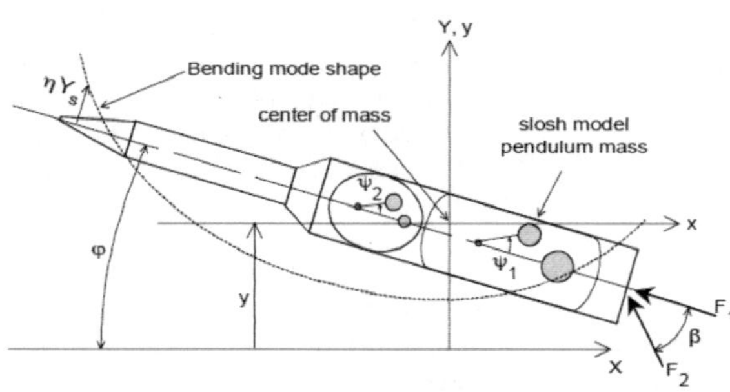

Figure 2: Mechanical Model Application for Spacecraft

2.2 DESCRIPTION OF MECHANICAL MODEL

The dynamic behaviour of liquid in a tank is represented by the mechanical model as shown in Figure 3. The liquid inside the tank is replaced by masses, springs, dampers and a disk.

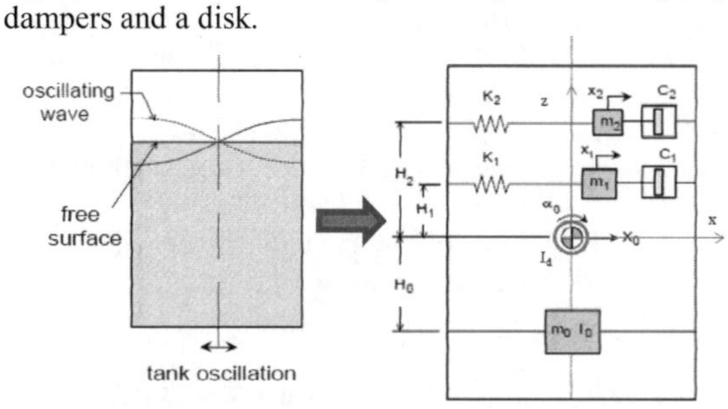

Figure 3: Spring-Mass Model

The fixed mass and mass inertia are represented by m_0 and I_0, respectively. Each slosh mode is represented by the spring-mass and damper system. Each mass including m_0 are located at H_n which is the height from the center of gravity of the liquid. Note that the positive direction of H_0 is opposite to the height of slosh masses. H_n is positive when the n-th mode mass is located above the COG, but H_0 is positive when it is located below the COG.

In this study, damping is neglected initially and added later. In addition, a disk I_d is introduced as the mass inertia of free rotating liquid which does not contribute to the mass inertia of liquid.

When the tank is excited at angular frequency ω by a small time-varying linear displacement X_0 along x axis and angular rotation α_0 about an axis through the COG, the spring masses deflect a distance x_n relative to the tank walls as a result of the tank motion. The amplitudes of horizontal force and moment exerted by X_0 and α_0 are obtained as follows;

$$\frac{F_{x\_amp}}{m_T\omega^2} = X_0\left\{1+\sum_{n=1}^{\infty}\frac{m_n}{m_T}\frac{\omega^2}{\omega_n^2-\omega^2}\right\} + h\alpha_0\left\{\sum_{n=1}^{\infty}\frac{m_n}{m_T}\frac{g}{h\omega_n^2}+\sum_{n=1}^{\infty}\frac{m_n}{m_T}\left(\frac{H_n}{h}+\frac{g}{h\omega_n^2}\right)\frac{\omega^2}{\omega_n^2-\omega^2}\right\} \quad (1)$$

$$\frac{M_{y\_amp}}{m_T h\omega^2} = X_0\left\{\sum_{n=1}^{\infty}\frac{m_n}{m_T}\frac{g}{h\omega_n^2}+\sum_{n=1}^{\infty}\frac{m_n}{m_T}\left(\frac{H_n}{h}+\frac{g}{h\omega_n^2}\right)\frac{\omega^2}{\omega_n^2-\omega^2}\right\} + h\alpha_0\left\{\begin{array}{l}\frac{I_y}{m_T h^2}+\sum_{n=1}^{\infty}\frac{m_n}{m_T}\frac{g}{h\omega_n^2}\frac{g}{h\omega^2} \\ +\sum_{n=1}^{\infty}\frac{m_n}{m_T}\frac{g}{h\omega_n^2}\left(\frac{2H_n}{h}+\frac{g}{h\omega_n^2}\right) \\ +\sum_{n=1}^{\infty}\frac{m_n}{m_T}\left(\frac{H_n}{h}+\frac{g}{h\omega_n^2}\right)^2\frac{\omega^2}{\omega_n^2-\omega^2}\end{array}\right\} \quad (2)$$

Here,

$$I_y = I_0 + m_0H_0^2 + \sum_{n=1}^{\infty} m_nH_n^2 = I_R - I_d \quad (3)$$

2.3 DETEMINATION OF MODEL PARAMETERS ANALYTICALLY

To determine the parameters of a mechanical model, horizontal force and moment shall be obtained analytically, numerically or experimentally.

For simple tank geometry such as a rectangular tank as shown in Figure 4, analytical solution can be obtained. Assuming the liquid in the tank is incompressible, no viscosity and irrotational, liquid motion can be described by velocity potential, Φ, which satisfies the following basic equations.

$$\nabla^2\Phi = \frac{\partial^2\Phi}{\partial x^2}+\frac{\partial^2\Phi}{\partial y^2}+\frac{\partial^2\Phi}{\partial z^2}=0 \quad (4)$$

$$\frac{\partial\Phi}{\partial t}+\frac{p}{\rho_c}+gz=0 \quad (5)$$

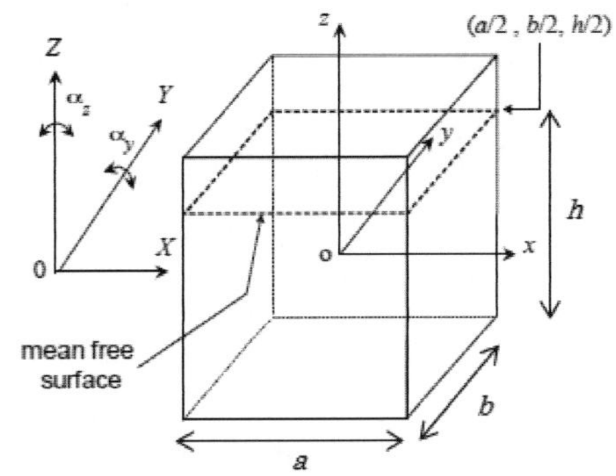

Figure 4: Rectangular Tank

When the tank is excited oscillatory at an angular frequency ω by a small time-varying linear displacement X_0 along x axis the boundary conditions are specified as follows;

$$\left[\frac{\partial^2\Phi}{\partial t^2}+g\frac{\partial\Phi}{\partial z}\right]_{z=\frac{h}{2}}=0 \text{ (at free surface)},$$
$$\left[\frac{\partial\Phi}{\partial z}\right]_{z=-\frac{h}{2}}=0 \text{ (at tank bottom)}, \quad (6)$$
$$\left[\frac{\partial\Phi}{\partial x}\right]_{x=\pm\frac{a}{2}}=X_0\omega\exp(i\omega t) \text{ (at tank wall)},$$
$$\left[\frac{\partial\Phi}{\partial y}\right]_{y=\pm\frac{b}{2}}=0 \text{ (at tank wall)}$$

First, by solving the eigenvalue problem, natural frequencies are determined as follows;

$$\omega_n^2 = g\lambda_n\tanh(\lambda_n h),\quad \lambda_n=\frac{(2n-1)\pi}{a} \quad (7)$$

Then, by solving the basic differential equation to satisfy the boundary conditions, the velocity potential is obtained as;

$$\Phi = X_0\omega\exp(i\omega t)\times \left\{x+\sum_{n=1}^{\infty}\frac{\omega^2}{\omega_n^2-\omega^2}\frac{4a(-1)^{n-1}}{\pi^2(2n-1)^2}\frac{\cosh[\lambda_n(z+h/2)]}{\cosh(\lambda_n h)}\sin(\lambda_n x)\right\} \quad (8)$$

The unsteady liquid pressure is;

$$p = -\rho \frac{\partial \Phi}{\partial t} = -i\rho X_0 \omega^2 \exp(i\omega t) \times$$
$$\left\{ x + \sum_{n=1}^{\infty} \frac{\omega^2}{\omega_n^2 - \omega^2} \frac{4a\,(-1)^{n-1}}{\pi^2 (2n-1)^2} \frac{\cosh\left[\lambda_n \left(z + \frac{h}{2}\right)\right]}{\cosh(\lambda_n h)} \sin(\lambda_n x) \right\} \tag{9}$$

The horizontal force and moment in the rectangular tank excited horizontally are obtained by integrating the liquid pressure over the tank wall area.

$$\frac{F_{x\_amp}}{m_T \omega^2} = X_0 \left\{ 1 + \frac{8a}{h} \sum_{n=1}^{\infty} \frac{\tanh(\lambda_n h)}{\pi^3 (2n-1)^3} \frac{\omega^2}{\omega_n^2 - \omega^2} \right\} \tag{10}$$

$$\frac{M_{y\_amp}}{m_T h \omega^2} = X_0 \left[\begin{array}{l} \frac{1}{12}\left(\frac{a}{h}\right)^2 \\ + \sum_{n=1}^{\infty} \left\{ \begin{array}{l} \frac{8\,(a/h)\tanh(\lambda_n h)}{\pi^3 (2n-1)^3} \times \\ \left[\frac{1}{2} - \frac{2\,(a/h)}{\pi(2n-1)} \tanh\left(\frac{\lambda_n h}{2}\right) + \frac{g}{h\omega_n^2} \right] \times \\ \frac{\omega^2}{\omega_n^2 - \omega^2} \end{array} \right\} \end{array} \right] \tag{11}$$

The horizontal force and moment in the rectangular tank excited by a small time-varying angular rotation α_0 about the axis through the COG can also be obtained by similar manner.

$$\frac{F_{x\_amp}}{m_T \omega^2} = h\alpha_0 \left[\begin{array}{l} \frac{1}{12}\left(\frac{a}{h}\right)^2 \\ + \sum_{n=1}^{\infty} \left\{ \begin{array}{l} \frac{8\,(a/h)\tanh(\lambda_n h)}{\pi^3 (2n-1)^3} \times \\ \left[\frac{1}{2} - \frac{2\,(a/h)}{\pi(2n-1)} \tanh\left(\frac{\lambda_n h}{2}\right) + \frac{g}{h\omega_n^2} \right] \times \\ \frac{\omega^2}{\omega_n^2 - \omega^2} \end{array} \right\} \end{array} \right] \tag{12}$$

$$\frac{M_{y\_amp}}{m_T h \omega^2} = h\alpha_0 \left[\begin{array}{l} \frac{I_y}{m_T h^2} + \frac{1}{12}\left(\frac{a}{h}\right)^2 \frac{g}{h\omega^2} \\ + \sum_{n=1}^{\infty} \left\{ \begin{array}{l} \frac{16\,(a/h)\tanh(\lambda_n h)}{\pi^3 (2n-1)^3} \frac{g}{h\omega_n^2} \times \\ \left(\frac{1}{2} - \frac{2\,(a/h)}{\pi(2n-1)} \tanh\left(\frac{\lambda_n h}{2}\right) + \frac{g}{2h\omega_n^2} \right) \end{array} \right\} \\ + \sum_{n=1}^{\infty} \left\{ \begin{array}{l} \frac{8\,(a/h)\tanh(\lambda_n h)}{\pi^3 (2n-1)^3} \times \\ \left[\frac{1}{2} - \frac{2\,(a/h)}{\pi(2n-1)} \tanh\left(\frac{\lambda_n h}{2}\right) + \frac{g}{h\omega_n^2} \right]^2 \times \\ \frac{\omega^2}{\omega_n^2 - \omega^2} \end{array} \right\} \end{array} \right] \tag{13}$$

$$I_y = \frac{1}{12} m_T \left(a^2 + h^2\right) \left\{ \begin{array}{l} 1 - \frac{4}{1 + (h/a)^2} \\ + \frac{768\, a/h}{\pi^5 \left[1 + (h/a)^2\right]} \sum_{n=1}^{\infty} \frac{\tanh\left[(2n-1)\pi h/2a\right]}{(2n-1)^5} \end{array} \right\} \tag{14}$$

The equation (1) shall be equal to the sum of the equations (10) and (12), and equation (2) shall be equal to the sum of the equations (11) and (13), and equation (3) shall be equal to equation (14). As a result, slosh mass, associated height and mass inertia of disk are determined as follows;

$$\frac{m_n}{m_T} = \frac{8a}{h} \frac{\tanh(\lambda_n h)}{\pi^3 (2n-1)^3} \tag{15}$$

$$\frac{H_n}{h} = \frac{1}{2} - \frac{2\,(a/h)}{\pi(2n-1)} \tanh\left(\frac{\lambda_n h}{2}\right) \tag{16}$$

$$I_R = \frac{1}{12} m_T \left(a^2 + h^2\right) \tag{17}$$

$$I_d = \frac{1}{12} m_T \left(a^2 + h^2\right) \left\{ \begin{array}{l} \frac{4}{1 + (h/a)^2} \\ - \frac{768\, a/h}{\pi^5 \left[1 + (h/a)^2\right]} \sum_{n=1}^{\infty} \frac{\tanh\left[(2n-1)\pi h/2a\right]}{(2n-1)^5} \end{array} \right\} \tag{18}$$

Remaining parameters, i.e., spring constant, fixed mass and height are determined as below;

$$K_n = m_n \omega_n^2 = m_T \frac{8g}{h} \frac{\tanh^2(\lambda_n h)}{\pi^2 (2n-1)^2} \tag{19}$$

$$m_0 = m_T - \sum_{n=1}^{\infty} m_n \tag{20}$$

$$H_0 = \sum_{n=1}^{\infty} \frac{m_n}{m_0} H_n \tag{21}$$

2.4 DETEMINATION OF MODEL PARAMETERS NUMERICALLY OR EXPERIMENTALLY

In many cases, determination by analytical solution is neither possible nor practical. For those cases, a model must be developed from numerical calculation or experimental measurement.

For practical application, higher order modes (n > 2) are usually of little concern, since the magnitudes of the slosh mass for those modes are very small compared to those of the fundamental mode. Only the first, lowest-frequency slosh mass needs to be considered.

The procedure for the determination of model parameter is to excite the tank by simple harmonic motion using both horizontal and rotating oscillation over a range of frequencies that encompasses the slosh fist natural frequency. And then we calculate or measure the resulting force and moment responses as a function of excitation frequency, and fit a model to the force and moment. The slosh mass, associated height and mass inertia of disk is calculated by following equations.

As shown in equations (22) to (24), slosh mass is determined by calculating or measuring the horizontal force against the horizontal oscillation. Slosh mass height is determined by the moment against the horizontal oscillation or by horizontal force against rotating oscillation. Mass inertia of free rotating liquid is determined by the moment against rotating oscillation. Those are summarized in Table 1.

$$m_1 = \left(\frac{F_{x\_amp}}{\omega^2 X_0} - m_T\right)\left(\frac{\omega_1^2 - \omega^2}{\omega^2}\right) \quad (22)$$

$$H_1 = \left(\frac{M_{y\_amp}}{m_1\omega^2 X_0} - \frac{g}{\omega_1^2}\right)\left(\frac{\omega_1^2 - \omega^2}{\omega^2}\right) - \frac{g}{\omega_1^2} \quad \text{or} \quad (23)$$

$$H_1 = \left(\frac{F_{x\_amp}}{m_1\omega^2 \alpha_0} - \frac{g}{\omega_1^2}\right)\left(\frac{\omega_1^2 - \omega^2}{\omega^2}\right) - \frac{g}{\omega_1^2}$$

$$I_d = I_R + \frac{m_1 g^2}{\omega_1^2\omega^2} + \frac{m_1 g}{\omega_1^2}\left(2H_1 + \frac{g}{\omega_1^2}\right) + m_1\left(H_1 + \frac{g}{\omega_1^2}\right)^2 \frac{\omega^2}{\omega_1^2 - \omega^2} - \frac{M_{y\_amp}}{\omega^2 \alpha_0} \quad (24)$$

Remaining parameters, spring constant, fixed mass and height are determined from the derived parameters as below;

$$K_1 = m_1\omega_1^2, \quad m_0 = m_T - m_1, \quad H_0 = \frac{m_1}{m_0} H_1 \quad (25)$$

Table 1: Determination of Model Parameters Numerically or Experimentally

Applied oscillation		Horizontal oscillation, X_0	
Calculation or Measurement		Hor'l force	Moment
Slosh mass	m_1	O	
Slosh mass height	H_1		O
Mass inertia of free rotating liquid	I_d		

Applied oscillation		Rotating oscillation, α_0	
Calculation or Measurement		Hor'l force	Moment
Slosh mass	m_1		
Slosh mass height	H_1	O	
Mass inertia of free rotating liquid	I_d		O

2.5 INCLUSION OF DAMPING

In order to avoid the violent resonance, equivalent linear viscous damping of liquid cargo is added.

The term, $\frac{\omega^2}{\omega_n^2 - \omega^2}$ is changed to $\frac{\omega^2}{\omega_n^2 - \omega^2 + i2\gamma_n\omega_n\omega}$.

2.6 VALIDATION OF MECHANICAL MODEL

The moment amplitude calculated by mechanical model is validated. The properties of rectangular tank with liquid cargo for the validation are shown in Table 2. The motion amplitudes for the test are 1.9 degrees, 3.8 degrees and 5.7 degrees.

These test conditions are the same as those of the experiments carried out by Bosch et al [14, 15].

The results are presented using a non-dimensional form. The non-dimensional moment is;

$$M_{nd\_amp} = \frac{M_{amp}}{\rho_c g B^3 L} \quad (26)$$

The reduced frequency is;

$$\sigma = \omega\sqrt{\frac{B}{g}} \quad (27)$$

Table 2: Tank Properties for Validation

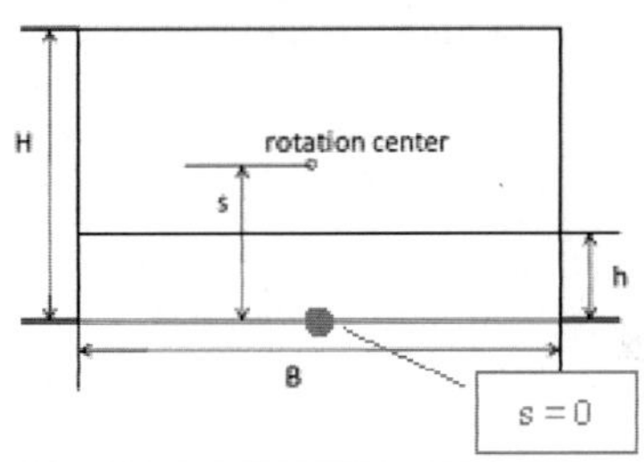

B	(m)	1.00
L	(m)	0.10
h	(m)	0.04
s	(m)	0.00
h/B		0.04
s/B		0.00

2.6 (a) Accuracy of Calculation

The analytical solution of sloshing in the rectangular tank is represented by the sum of the infinite series; however, only the first, i.e., the lowest, frequency slosh mass can practically be used when the mechanical model parameters are derived by other than analytical solution.

It is commonly accepted that higher order modes (n > 2) can usually be ignored since the magnitudes of the slosh mass for those modes are very small compared to the fundamental mode. Table 3 shows this relationship.

Table 3: Accuracy of Calculation for Cut-Off Mode $\alpha_0 = 1.9$ degrees, $\gamma = 0.1\%$

	Reduced frequency σ				
	0.032	0.255	0.511	0.766	0.990
n = 1	0.9856	0.9876	0.9946	1.0095	1.0366
n = 5	0.9998	0.9999	0.9999	1.0001	1.0003
n = 20	1.0000	1.0000	1.0000	1.0000	1.0000
n = 100	1.0000	1.0000	1.0000	1.0000	1.0000

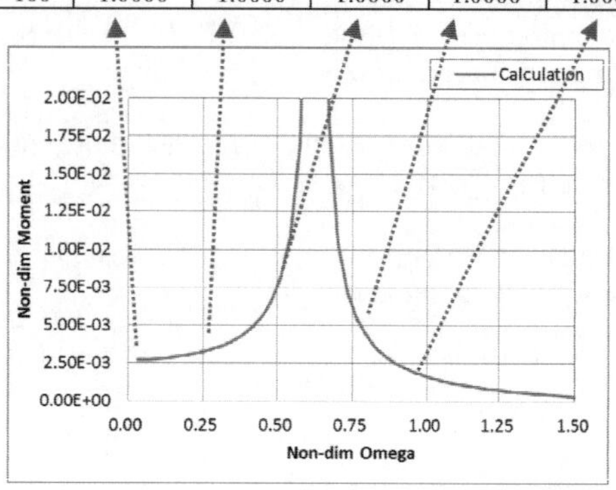

Table 3 shows the ratio of the calculated moment amplitude adding up to n modes to the calculated moment considering 100 modes. The convergence is quick and the summation up to 100 modes is sufficiently accurate as the analytical solution. The error of calculated moment amplitude considering only the 1st mode (n=1) in the range from quasi-static region to resonant region is within 2%. In high frequency region, the accuracy becomes worse; however, the moment in this frequency region is small and not so important for practical application.

2.6 (b) Comparison to Experiment

Comparisons of the amplitude of the moment on the tank by liquid motion are shown in Figure 5. Experimental results used for the comparison are taken from the graphs in Refs. [14, 15].

As shown in the figure, the numerical results agree fairly well with the experimental data except in the vicinity of the natural frequency.

As explained in Refs. [14, 15], free surface effect dominates the magnitude of the moment in the very low frequencies. For somewhat larger frequencies, low standing wave increases the magnitude of the moment. The mechanical model represents the liquid motion fairly well.

With an increase in the frequency, the non-linear liquid motion such as a train of small waves, hydraulic jump, etc. appears and the amplitude of the moment does not increase significantly.

However, the amplitude of the moment calculated by the mechanical model increases further until the extremely large peak at the natural frequency due to very small damping.

At the end of this frequency range, the hydraulic jump degenerates into a solitary wave with steep sides. Consequently the mass transport becomes less and so the amplitude of the moment drops. The mechanical model representation shows good agreement with the measured data again.

3. COUPLED EQUATION OF FLOATING BODY MOTION AND LIQUID CARGO EFFECTS

3.1 FLOATING BODY MOTION WITHOUT LIQUID CARGO

The floating body motion relative to COG of the floating body is described as the following equation in frequency domain.

$$\left[-\omega^2\left(\mathbf{M}_G + \mathbf{M}_{A\_G}\right) + i\omega\,\mathbf{B}_G + \mathbf{C}_G\right]\cdot \mathbf{X}_G = \mathbf{F}_G^{DI} \qquad (28)$$

where $\mathbf{X}_G$ is floating body motion matrix, $\mathbf{M}_G$ is mass matrix of the floating body, $\mathbf{M}_{A\_G}$ is hydrodynamic added mass matrix, $\mathbf{B}_G$ is hydrodynamic potential and quadratic damping matrix, $\mathbf{C}_G$ is hydrostatic restoring matrix, and $\mathbf{F}_G^{DI}$ is hydrodynamic excitation force matrix due to incoming wave and diffraction wave.

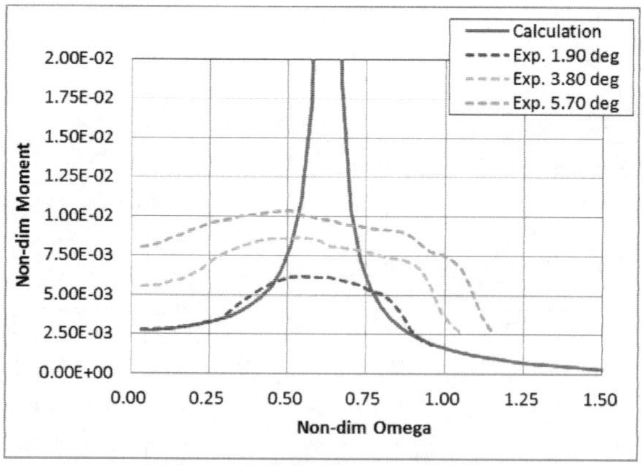

(a) $\alpha_0 = 1.9$ degrees, $\gamma = 0.1\%$

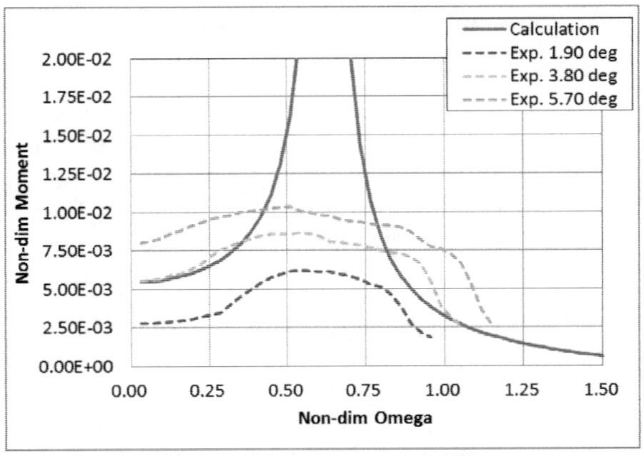

(b) $\alpha_0 = 3.8$ degrees, $\gamma = 0.1\%$

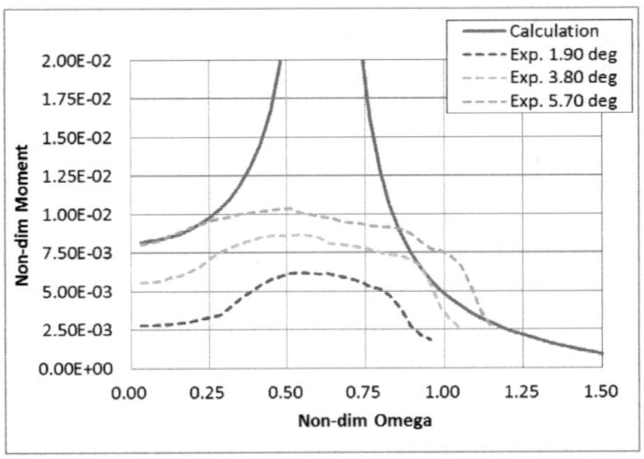

(c) $\alpha_0 = 5.7$ degrees, $\gamma = 0.1\%$

Figure 5: Comparisons of the moments

3.2 COUPLED FLOATING BODY MOTION WITH LIQUID CARGO

To obtain the coupled motion equation of floating body and liquid cargo effects, the forces induced by liquid cargo are added to the equation of the floating body motion. When there are multiple tanks, all tank liquid forces are summed and added. Note that the mass matrix, $\mathbf{M_{B\_G}}$ shall exclude the liquid mass, but mass inertia shall be calculated relative to COG of the floating body including liquid cargo.

$$\left[-\omega^2\left(\mathbf{M}_{\mathbf{B\_G}}+\mathbf{M}_{\mathbf{A\_G}}\right)+i\omega\,\mathbf{B}_{\mathbf{G}}+\mathbf{C}_{\mathbf{G}}\right]\cdot\mathbf{X}_{\mathbf{G}} = \mathbf{F}_{\mathbf{G}}^{\mathbf{DI}}+\sum\left[\mathbf{F}_{\mathbf{T}}^{\mathbf{L}}+\mathbf{F}_{\mathbf{TG}}\right]\cdot\mathbf{X}_{\mathbf{G}} \tag{29}$$

$\mathbf{M_{B\_G}}$ is mass matrix of the floating body (excluding liquid cargo) relative to COG of the floating body (including liquid cargo)

$$\mathbf{M}_{\mathbf{B\_G}} = \begin{pmatrix} M_B & 0 & 0 & m_{14} & m_{15} & m_{16} \\ 0 & M_B & 0 & m_{24} & m_{25} & m_{26} \\ 0 & 0 & M_B & m_{34} & m_{35} & m_{36} \\ m_{41} & m_{42} & m_{43} & I_{11}^{G} & I_{12}^{G} & I_{13}^{G} \\ m_{51} & m_{52} & m_{53} & I_{21}^{G} & I_{22}^{G} & I_{23}^{G} \\ m_{61} & m_{62} & m_{63} & I_{31}^{G} & I_{23}^{G} & I_{33}^{G} \end{pmatrix}$$

Here,

$$m_{14} = m_{41} = 0,\quad m_{15} = m_{51} = M_B(z_{BG} - z_G),\quad m_{16} = m_{61} = -M_B(y_{BG} - y_G)$$
$$m_{24} = m_{42} = -M_B(z_{BG} - z_G),\quad m_{15} = m_{51} = 0,\quad m_{16} = m_{61} = M_B(x_{BG} - x_G)$$
$$m_{16} = m_{61} = M_B(y_{BG} - y_G),\quad m_{15} = m_{51} = -M_B(x_{BG} - x_G),\quad m_{16} = m_{61} = 0$$

$$I_{11}^{G} = I_{11}^{BG} + M_B\left[(y_{BG} - y_G)^2 + (z_{BG} - z_G)^2\right]$$
$$I_{22}^{G} = I_{22}^{BG} + M_B\left[(x_{BG} - x_G)^2 + (z_{BG} - z_G)^2\right]$$
$$I_{33}^{G} = I_{33}^{BG} + M_B\left[(x_{BG} - x_G)^2 + (y_{BG} - y_G)^2\right]$$

$$I_{12}^{G} = I_{21}^{G} = I_{12}^{BG} - M_B\left[(x_{BG} - x_G)(y_{BG} - y_G)\right]$$
$$I_{13}^{G} = I_{31}^{G} = I_{13}^{BG} - M_B\left[(x_{BG} - x_G)(z_{BG} - z_G)\right]$$
$$I_{23}^{G} = I_{32}^{G} = I_{23}^{BG} - M_B\left[(y_{BG} - y_G)(z_{BG} - z_G)\right]$$

$\mathbf{F_T}^{\mathbf{L}}$ is tank liquid force matrix relative to local tank origin.

$$\mathbf{F}_{\mathbf{T}}^{\mathbf{L}} = m_T\omega^2 \times \begin{pmatrix} 1+A_x & 0 & 0 & 0 & B_x & 0 \\ 0 & 1+A_y & 0 & -B_y & 0 & 0 \\ 0 & 0 & 1 & 0 & 0 & 0 \\ 0 & -B_y & 0 & \dfrac{I_x}{m_T}+C_y+D_y\dfrac{1}{\omega^2} & 0 & 0 \\ B_x & 0 & 0 & 0 & \dfrac{I_y}{m_T}+C_x+D_x\dfrac{1}{\omega^2} & 0 \\ 0 & 0 & 0 & 0 & 0 & 0 \end{pmatrix}$$

Where,

$$A_x = \sum_{n=1}^{\infty}\frac{m_{xn}}{m_T}\frac{\omega^2}{\omega_{xn}^2 - \omega^2}$$
$$A_y = \sum_{n=1}^{\infty}\frac{m_{yn}}{m_T}\frac{\omega^2}{\omega_{yn}^2 - \omega^2}$$

$$B_x = \sum_{n=1}^{\infty}\frac{m_{xn}}{m_T}\frac{g}{\omega_{xn}^2} + \sum_{n=1}^{\infty}\frac{m_{xn}}{m_T}\left(H_{xn} + \frac{g}{\omega_{xn}^2}\right)\frac{\omega^2}{\omega_{xn}^2 - \omega^2}$$
$$B_y = \sum_{n=1}^{\infty}\frac{m_{yn}}{m_T}\frac{g}{\omega_{yn}^2} + \sum_{n=1}^{\infty}\frac{m_{yn}}{m_T}\left(H_{yn} + \frac{g}{\omega_{yn}^2}\right)\frac{\omega^2}{\omega_{yn}^2 - \omega^2}$$

$$C_x = \sum_{n=1}^{\infty}\frac{m_{xn}}{m_T}\frac{g}{\omega_{xn}^2}\left(2H_{xn} + \frac{g}{\omega_{xn}^2}\right) + \sum_{n=1}^{\infty}\frac{m_{xn}}{m_T}\left(H_{xn} + \frac{g}{\omega_{xn}^2}\right)^2\frac{\omega^2}{\omega_{xn}^2 - \omega^2}$$
$$C_y = \sum_{n=1}^{\infty}\frac{m_{yn}}{m_T}\frac{g}{\omega_{yn}^2}\left(2H_{yn} + \frac{g}{\omega_{yn}^2}\right) + \sum_{n=1}^{\infty}\frac{m_{yn}}{m_T}\left(H_{yn} + \frac{g}{\omega_{yn}^2}\right)^2\frac{\omega^2}{\omega_{yn}^2 - \omega^2}$$

$$D_x = \sum_{n=1}^{\infty}\frac{m_{xn}}{m_T}\frac{g}{\omega_{xn}^2}$$

$$D_y = \sum_{n=1}^{\infty} \frac{m_{yn}}{m_T} \frac{g}{\omega_{yn}^2}$$

$$I_y = I_{yR} - I_{yd}$$
$$I_x = I_{xR} - I_{xd}$$

$\mathbf{F_{TG}}$ is coordinate transfer matrix of the tank liquid force from local tank origin to COG.

$$\mathbf{F_{TG}} = \begin{bmatrix} 0 & -\mathbf{F_T^{L11}V} \\ \mathbf{VF_T^{L11}} & \mathbf{VF_T^{L12}} - \mathbf{F_T^{L21}V} - \mathbf{VF_T^{L11}V} \end{bmatrix}$$

Where,

$$\mathbf{F_T^L} = \begin{bmatrix} \mathbf{F_T^{L11}} & \mathbf{F_T^{L12}} \\ \mathbf{F_T^{L21}} & \mathbf{F_T^{L22}} \end{bmatrix}$$

$$\mathbf{F_T^{L11}} = m_T\omega^2 \times \begin{pmatrix} 1+A_x & 0 & 0 \\ 0 & 1+A_y & 0 \\ 0 & 0 & 1 \end{pmatrix} \qquad \mathbf{F_T^{L12}} = m_T\omega^2 \times \begin{pmatrix} 0 & B_x & 0 \\ -B_y & 0 & 0 \\ 0 & 0 & 0 \end{pmatrix}$$

$$\mathbf{F_T^{L21}} = m_T\omega^2 \times \begin{pmatrix} 0 & -B_y & 0 \\ B_x & 0 & 0 \\ 0 & 0 & 0 \end{pmatrix}$$

$$\mathbf{F_T^{L22}} = m_T\omega^2 \times \begin{pmatrix} \frac{I_x}{m_T} + C_y + D_y\frac{1}{\omega^2} & 0 & 0 \\ 0 & \frac{I_y}{m_T} + C_x + D_x\frac{1}{\omega^2} & 0 \\ 0 & 0 & 0 \end{pmatrix}$$

$$\mathbf{V} = \begin{bmatrix} 0 & -(z_{TG} - z_G) & y_{TG} - y_G \\ z_{TG} - z_G & 0 & -(x_{TG} - x_G) \\ -(y_{TG} - y_G) & x_{TG} - x_G & 0 \end{bmatrix}$$

The forces induced by liquid cargo are finally divided into added mass part and hydrostatic restoring part.

$$\mathbf{F_T^L} + \mathbf{F_{TG}} = \omega^2 \mathbf{M_{LA\_G}} + \mathbf{C_{L\_G}} \tag{30}$$

And the equation of floating body motion coupled with the liquid cargo effects is obtained as follows;

$$\begin{bmatrix} -\omega^2 \left(\mathbf{M_{B\_G}} + \mathbf{M_{A\_G}} + \sum \mathbf{M_{LA\_G}}\right) \\ + i\omega \mathbf{B_G} + \mathbf{C_G} - \sum \mathbf{C_{L\_G}} \end{bmatrix} \cdot \mathbf{X_G} = \mathbf{F_G^{DI}} \tag{31}$$

4. NUMERICAL DEMONSTRATIONS

In order to demonstrate the capabilities of the present method, added mass coefficients for sway and roll due to the liquid cargo, and RAO of roll motion with and without the liquid cargo effect are calculated and compared with the numerical results obtained by Malenica et al [3].

Malenica et al chose the example from Molin et al [4]. It is a rectangular barge (length L=3.0m, width B=1.0m, draught T=0.108m) with two rectangular tanks (length L=0.8m, width B=0.25m, draught T=0.19m) located on the barge deck (bottom of the tanks at 0.3m).

They showed the added masses coefficients for sway and roll. The figure is reproduced here in Figure 6. Dotted line, equal to 1, represents the results for the case when the liquid in the tanks is treated as rigid. The sloshing resonant frequency is clearly identified and damping effect is observed.

The added mass coefficients for sway and roll calculated by the mechanical model are shown in Figure 7. The added mass coefficients are calculated considering only the first, i.e., the lowest, frequency slosh mass.

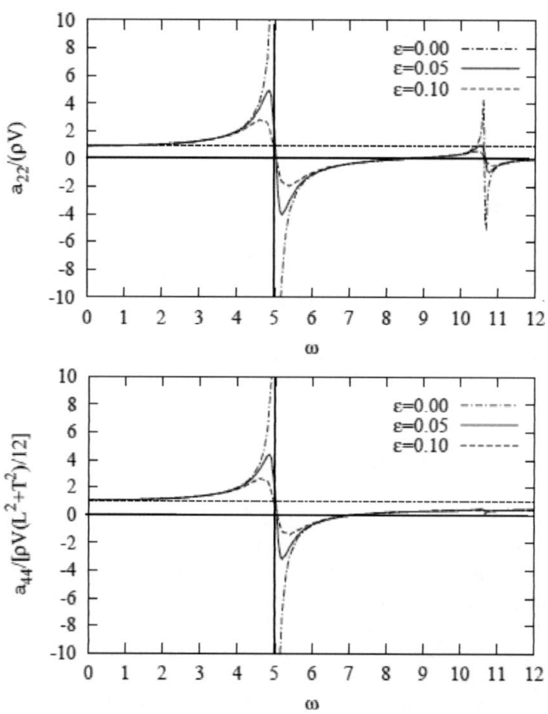

Figure 6: Added Mass Coefficients for Sway and Roll by Malenica et al [3]

The RAO of roll motion in beam sea calculated by Malenica et al is reproduced in Figure 8. When the liquid in the tanks is treated as rigid, one peak roll RAO is observed. When the liquid motion in tanks is considered, two peaks appeared.

The RAO of roll motion in beam sea coupled with the liquid cargo forces calculated by the mechanical model is obtained.

The hydrodynamic coefficients for rectangular barge are calculated by ANSYS AQWA [16].

As shown in Figure 9, calculated roll RAO also catches the characteristics of liquid cargo effect to the barge roll motion.

Now authors conclude that the mechanical model is operational and can be used in common studies of the FLNG with liquid cargo in tanks.

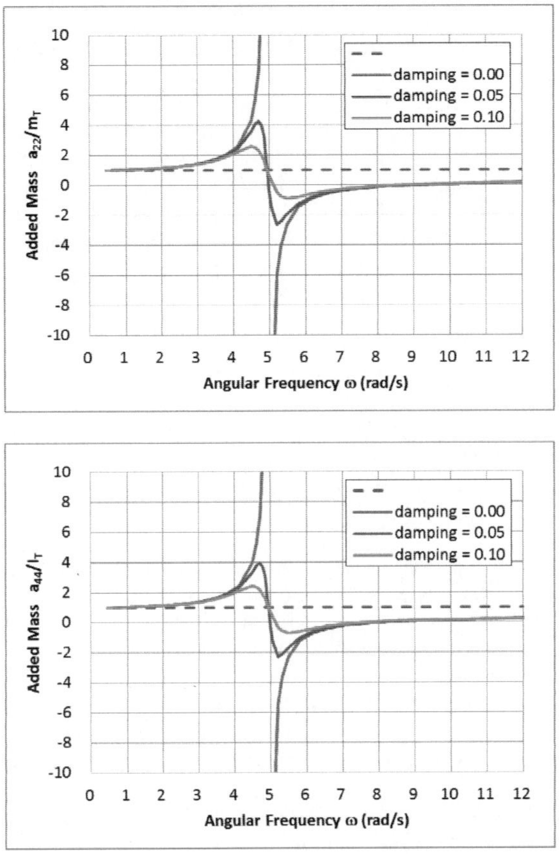

Figure 7: Added Mass Coefficients for Sway and Roll

5. LIMITATIONS

The mechanical model to represent the liquid cargo behaviour is a simple and efficient analytical model, which can be a useful tool for marine and offshore industry to utilize in the early design stage. However, due to the nature of the mechanical model, the following liquid behaviour is not well represented.

- Non-linear effects such as traveling wave, hydraulic jump, breaking wave, vertical sloshing, etc.
- 3D effects such as swirling, yaw moment, etc.

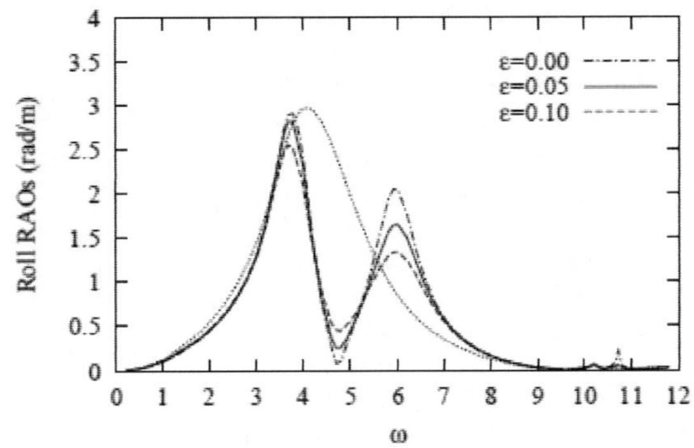

Figure 8: RAO of Roll motion in Beam Sea by Malenica et al [3]

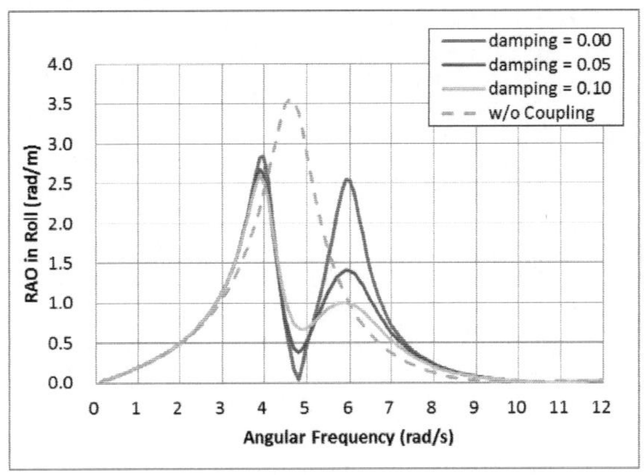

Figure 9: RAO of Roll motion in Beam Sea

6. CONCLUSIONS

An efficient method to estimate the dynamic coupling between the liquid cargo and the floating body motion is presented in this paper.

- A simple and less computationally demanding motion analysis method that can consider liquid sloshing is required in practical application for FLNG development.
- The mechanical model to represent the liquid cargo effects is a suitable and proven solution.
- The liquid cargo effects are represented by the spring-mass type mechanical model and they are coupled with floating body motion.
- The parameter of mechanical model can be determined analytically, numerically or experimentally.
- The method is validated against the results from the literature.
- Limitation of the method is noted.

The development of the method has just started. Authors plan to continue the investigation and development further. The investigation plans at this moment are as follows;

- Validation of the force and the moment caused by the liquid motion when the tank is in the event of the high liquid level condition,
- Validation of parameter determination method using numerical or experimental result,
- Investigation of the detailed characteristics of coupled floating body motion, and
- Case studies using the proposed method.

7. ACKNOWLEDGEMENTS

The authors appreciate the following research partners and colleagues for their kind support and advise; Prof. K. Nishimoto (University of São Paulo/USP), Dr. L.Y. Cheng (USP), Dr. X. Wang (Newcastle University), Mr. T. Kawahashi (Yokohama National University/YNU), Ms. R. Hata (YNU) and Mr. G.M. Karuka (YNU). Also special thanks to JGC Corporation and colleagues in JGC for giving opportunity and permission to publish this paper.

8. REFERENCES

1. FALTINSEN, O.M. and TIMOKHA, A.N., 'Sloshing', *Cambridge University Press*, 2009.

2. ROGNEBAKKE. O. F. and FALTINSEN, O. M., 'Coupling of sloshing and ship motions', *Journal of Ship Research, 47(3), 208–221*, 2003.

3. MALENICA, Š., ZALAR, M. & CHEN, X.B., 'Dynamic coupling of seakeeping and sloshing', *13th ISOPE Conference 23–28, Honolulu, USA*, 2003.

4. MOLIN, B., REMY, F., RIGAUD, S., and DE JOUETTE, C., 'LNG-FPSO's: frequency domain, coupled analysis of support and liquid cargo motions' *Proceedings of IMAM conference*, 2002.

5. WANG, X., 'Coupled Analysis of Ship Motions and Tank Sloshing of LNG Carriers – Study of a Time-domain Numerical Method and its Application', *PhD thesis, Yokohama National University*, 2012.

6. WANG, X. and ARAI, M., 'Research on computational method of coupled ship motions and sloshing', *Journal of the Japan Society of Naval Architects and Ocean Engineers, 14, 97–104*, 2011.

7. YAMASHITA, S., 'Motions of a floating body with a liquid storage tank', *Transactions of the West-Japan Society of Naval Architects, 61, 155–165*, 1981.

8. ROCHA, T. P., NISHIMOTO, K., et al. 'Experimental Investigation on the Influence of Liquid Cargo in Floating Vessels Motions', *OTC-26203-MS*, 2015.

9. LEE, S. J., 'The effects of LNG-sloshing on the global responses of LNG-carriers', *PhD thesis, Texas A&M University*, 2008.

10. LEE, S. J., KIM, M. H., LEE, D. H., KIM, J. W., and KIM, Y. H., 'The effects of LNG tank sloshing on the global motions of LNG carriers', *Ocean Engineering, 34, 10–20*, 2007.

11. ORIMO, Y., DE WILDE, J., ICHIMARU, Y., TERASHIM, T., VAN DEN BERG, J., 'Methodology to Determine Floating LNG Tank Capacity by Combination of Side-by-Side Down-Time Simulation and Cost/Benefit Analysis', *OTC 23019, 483–492,* 2012.

12. DODGE, F. T., 'The New "Dynamic Behavior of Liquids in Moving Containers"', *Southwest Research Institute*, 2000.

13. KOMATSU, K., 'Sloshing – Liquid Motion and Tank Shell Vibration –', 2015. (in Japanese)

14. Van Den Bosch, J.J. and Vugts, J.H., 'Roll Damping by Free-Surface Tanks', *Shipbuilding Laboratory of the Technological University, Delft, Report No. 134*, 1965.

15. Van Den Bosch, J.J. and Vugts, J.H., 'On Roll Damping by Free-Surface Tanks', *RINA,* 1966.

16. ANSYS Aqwa Theory Manual, USA, 2013.

9. AUTHORS BIOGRAPHY

Akira Nakashima holds the current position of Civil and Structural Engineer at JGC Corporation and PhD student at Yokohama National University. He is working on research of FLNG motion affected by internal tank sloshing and gap water between FLNG and LNG carrier in university.

Makoto Arai is Professor at the Department of Systems Design for Ocean-Space, Faculty of Engineering, Yokohama National University. His main area of interests includes the numerical modelling of sloshing, maritime logistics, ballast-free ship, etc.

LNG CARRIERS AND PORT DESIGN

A.P. Crowle, CB&I, United Kingdom

SUMMARY

This paper describes the port design requirements for LNG import and export terminals and their specific interface with LNG carriers. The berthing and mooring requirements for the largest and smallest spherical and membrane containment systems LNG carriers will be included. The berthing analyses methods for the design range of LNG carriers produce berthing energies at ballast draft and loaded draft conditions and the effect of these loadings on fender and berthing dolphin design are discussed.

The design interface between piping, navigation aids, lighting, jetty control building and the LNG carrier are reported. Safety issues discussed include the size of impounding basins, fire-fighting equipment, minimum distance between LNG carriers at the berths, location of emergency shut down valves, height of jetty lightning protection, and emergency egress from a jetty berth. Interaction between equipment such as loading arms, access gangways, mooring hooks, and berth monitoring equipment are described.

The methods of assessing LNG carrier manoeuvring simulations with regards to the viability and possible limitations of the navigation channel width, water depth at the berths and the berthing and un-berthing manoeuvres at the LNG Terminal are reported. The paper discusses the factors to be considered with respect to loading/unloading LNG onto LNG carriers to determine the baseline berth utilisation rate. Furthermore, this paper examines the potential down time for marine operations of LNG carrier arrival, loading/unloading and departure due to local weather conditions such as visibility, wind, waves, current and sea ice.

NOMENCLATURE

[Symbol]	[Definition] [(unit)]
COG	Centre of Gravity
CUM	Cubic metres
FSRU	Floating, Storage, Regas Unit
GM	Metacentre above COG
LOA	Length overall
LNG	Liquefied Natural Gas
MOF	Material offloading facility
MTPA	Million tonnes per annum
QRH	Quick release hook
SPMT	Self-propelled modular transporter
T	Tonnes

1. INTRODUCTION

LNG terminal facilities require careful planning and analysis with regard to technical requirements, investment costs, economic incentives, process requirements, capital investment, and other special requirements of the Government and local interest groups.
A typical 2 berth LNG terminal is shown in Figure 1.

These site selection and LNG jetty design issues would include:

- Water depth, at chart datum, based on a common draft for medium to large LNG carriers of about 12m
  - 14m in sheltered locations
  - 16m in exposed locations
- Throughput of LNG
  - Probabilistic simulations and risk analysis
  - Evaluation of operational scenarios
  - Evaluation of trade-offs between LNG terminal downtime vs. infrastructure improvements
- Berth optimisation studies (number, type, size) based on throughput in million tonnes per annum (MTPA)
  - Less than 10 MTPA 1 berth is usually sufficient
  - From 10 to 20 MTPA 2 berths are required
- Navigation studies, which will show
  - Whether dredging is required
  - Location of navigation aids
- Site data characterisation
  - Bathymetry
  - Topographic survey data
  - Geotechnical, geophysical & seismic investigation
  - Metocean data
  - Ice data
- Berth orientation optimisation depends on
  - Predominant wave direction
  - Most likely wind direction
- Berth Layout
  - Platform and dolphin spacing/configuration
  - Support craft facilities
  - Service and maintenance vehicle requirements
  - Access and traffic flow
  - Secondary egress considerations
- Equipment requirements
  - Safety (fire water pump, fire monitors)
  - LNG loading/unloading arms
- Support services evaluation
  - Fire response capability
  - Fire water sourcing
  - Tug and line boat availability
- Berth availability evaluation

  - Considering metocean data
  - Average size of LNG carrier at the site
  - Marine traffic
  - Maintenance considerations
  - LNG storage capacity
- Design life, is generally 25 to 30 years from start-up of the LNG plant
- Design and construction method, the jetty substructure can be constructed by
  - Walk out method from the shore
  - Construction jack ups
  - Floating crane
  - Or a combination of methods
- Security risk planning includes
  - Fencing
  - Security cameras
  - Radar
- Cost estimating for
  - Soil investigations
  - Metocean monitoring
  - Design
  - Procurement
  - Jetty substructure
  - Topside equipment
  - Piping and valves
  - Electrical and instruments
  - Safety equipment
  - Tugs and line handling boats during operations
- Design, procurement and construction scheduling requires the
  - Pioneer Dock to be ready to import materials for the temporary labour camp
  - Material Offloading Facility (MOF) ready for import of equipment and modules
  - LNG berth ready prior to testing onshore plant

2. HISTORY OF LNG PORTS IN THE UK

2.1 CANVEY ISLAND

Due to its location and the deep water mooring, Canvey Island, on the Thames Estuary in Essex, UK, was chosen to be the site of a pilot scheme, the first LNG import terminal in the world. LNG was transported by sea from Lake Charles, Louisiana to purpose-built tanks at the Canvey site, UK. The construction of the pilot scheme was completed in 1959 when the The *Methane Pioneer*, a converted World War II liberty freighter and the first LNG carrier. During 1959, seven cargoes of LNG were transported to the new Canvey Terminal. The experiment to transport LNG by sea, had been a success and with government approval, a contract with Algeria was made, to import LNG into the U.K. for industrial and commercial use.

In 1962, work started to construct the permanent LNG terminal which consisted of six double-skinned storage tanks, a purpose-built jetty and evaporators. Two purpose-built LNG carriers were commissioned to carry the LNG from Algeria to Canvey, these being *Methane Princess* and *Methane Progress;* the first being built at Vicker's Yard at Barrow-in-Furness, UK, and the second at Harland and Wolfe Shipyard in Belfast, UK. By 1964, the work was completed. The Canvey Island site no longer imports LNG.

1.1 2.2South Hook LNG terminal

Close to Milford Haven in UK, is the South Hook LNG regasification terminal which is the largest LNG terminal in Europe. The LNG import terminal opened in March 2009, and the jetty can moor two Q-MAX LNG carriers (see Table 1) at the same time, although unloading only takes place from one of them. Each unloading berth is equipped with:

- 3 unloading arms
- 1 hybrid arm
- 1 vapour return arm
- 1 tower access gangway
- 1 column access gangway
- 1 drain drum
- 2 lightning protection masts
- 2 fire monitor towers

2.3 DRAGON LNG

Dragon LNG is also located on Milford Haven and has one unloading berth. First gas to the Dragon facility was delivered in July 2009. The unloading berth is equipped with:

- 2 unloading arms
- 1 hybrid arm
- 1 vapour return arm
- 1 tower access gangway

2.4 GRAIN LNG

With two berths on the Medway in Kent, England, Grain LNG has the ability to unload, reload and tranship LNG. Each unloading berth is equipped with:

- 2 unloading arms
- 1 hybrid arm
- 1 vapour return arm
- 1 tower access gangway

2.5 TEESSIDE

Teesside GasPort, UK, was the world's first dockside floating regasification facility located near Middlesbrough in the United Kingdom. Placed in service in February 2007, it delivered natural gas into the country's National Transmission System. The Teesside GasPort was brought into service within 12 months of site selection and was capable of accommodating a 150,900 m^3 Floating Storage ReGas Unit (FSRU). The project was decommissioned in 2015.

3. SITE IDENTIFICATION

3.1 NON TECHNICAL REQUIREMENTS

The site identification is driven by assessing a number of selection criteria. Once a site has been chosen, a rigorous multi-discipline review is made considering construction and operation issues. Some of the site identification issues which may impact project success include:

- Fully understanding national and local government preferences in order to minimize approval time
- Distance from the gas reservoir to onshore site for a LNG liquefaction terminal
- Distance from main gas distribution pipelines for a LNG import terminal
- Land ownership
- Environmental Constraints of flora and fauna
- Need for a deep water harbour and unrestricted LNG Carrier access
- Impact of site works and material availability on project cost and schedule
- Access to supporting infrastructure (e.g. Airport, rail)

3.2 SIZE AND TYPE OF LNG CARRIER

The size and type of the LNG carriers expected at the LNG berth are key criteria in site identification as they effect the number of LNG berths, LNG loading/unloading arms and the location of the breasting and mooring dolphins. In addition the position of the LNG carrier manifold has an influence on the location of the dolphins relative to amidships.

Spherical containment system LNG carriers typically have a higher wind area than equivalent volume membrane containment system LNG carriers. The centre of the manifold on a spherical containment system LNG carrier is typically offset from amidships up to 30m. Whereas the centre of the Manifold on membrane containment system LNG carrier is close to amidships.

The number and location of mooring dolphins and breasting dolphins is a function of:

- LNG carrier size
- Manifold location
- Mooring loads, partly a function of wind area
- Berthing speed and hence berthing load

The water depth at the mooring and breasting dolphins is a function of the LNG carrier draft, which is typically about 12m, plus an allowance for heaving in waves. Thus for a sheltered location a minimum water depth at extreme low tide is 14 m and in an exposed location about 16 m.

3.3 LNG JETTY LAYOUT

The jetty orientation is determined based on wave modelling data and is usually selected to keep the highest recurrence of the higher energy waves as close to directly on the bow of a moored LNG carrier as possible.

The access channel to the LNG berth will typically have water depths of at least 20 metres (with reference to Chart Datum). The width of the access channel should be at least three times the width of the largest LNG carrier and should be as straight as possible or with gently curving corners.

The manoeuvring area in the vicinity of the Jetty Loading Platform will utilize the naturally occurring water depths or require dredging. A turning circle with a diameter of approximately two times the length of the largest design of LNG carrier fits well within the manoeuvring area adjacent to the jetty loading platforms. The manoeuvring diameter may be reduced in very sheltered locations.

3.4 LOADING PLATFORMS

The jetty loading platforms, or unloading platforms, provide space for all topsides equipment required for the LNG loading operations. This includes features such as the marine LNG loading/unloading arms, piping, valves and other equipment, a tower gangway access structure, a service crane, vehicular manoeuvring and parking, firefighting equipment, area lighting, LNG containment and general drainage.

See Figure 2 for a typical loading platform

The jetty loading platform is usually a pile supported structure, but could be a gravity based structure. The piles of the unloading platform are located to maintain at least one metre clearance on the berthing LNG carrier hull when the LNG carrier is subject to three degrees (3°) roll under the maximum compression of the fenders. The elevation of the jetty loading platforms is established in a manner to allow the crests of the design waves to pass clearly under the structural elements of the jetty loading platform main deck to prevent uplift loads from being imposed on the structure. All structural steel elements of the jetty loading platforms must be adequately coated for corrosion protection to the highest marine standards. Any steel that may be subject to LNG spillage needs either a low temperature epoxy coating or a concrete structure could be used instead.

The liquid arms (two or three), one vapour return arm and one hybrid arm (primary liquid, secondary vapour) are installed on the jetty loading platform to accommodate loading the full range of design LNG carriers in the full range of operating conditions. The centre of the vapour return arm is considered to be the centre of the loading platform. The centre line of the LNG loading/unloading arm risers are located about 3.5 metres back from the edge of the jetty loading platform and they are spaced about 4.0 metres centre to centre. Access stairs and elevated steel platforms are provided to allow personnel access to the front of the marine loading arms for servicing the triple swivel and emergency release coupling mechanisms and the back of the marine loading arms for access to nitrogen valves. The elevated steel platform also accommodates dummy manifolds for each marine loading arm. The jetty loading platform accommodates all necessary piping, valves and equipment to support the LNG arm operation. A tower gangway access structure is provided on the jetty loading platform for safe access to and from the LNG carrier.

The tower gangway access structure must be designed to accommodate the deck landing areas for the full range of design LNG carriers through the full range of operating conditions. The tower gangway access structure is located to allow access to the main deck of the LNG carrier aft of the manifold area in the typical mooring arrangement (port-side-to). The gangway landing area on board the LNG carrier is in the range of 15 to 27 metres aft of the vapour line. The gangway stair unit at the LNG carrier landing end is a tripod type with rollers, with stairs able to swing either to forward or aft and a deck footprint of approximately 2 metres by 1 metre.

A service crane may be integrated with the tower gangway access structure. The service crane should have a hoisting capacity of about five metric tonnes to allow the safe transfer of materials from the jetty loading platform to the main deck of the LNG carrier.

Firefighting equipment, including fire monitors and dry chemical systems are required on the jetty loading platform.

Containment kerbs of sufficient height are provided around the marine loading arms and all process piping that carries LNG on the jetty loading platform. The LNG containment area on the jetty loading platform is sloped in such a manner so that liquids flow toward the adjacent impounding basins. Containment kerbs should also be provided around all oil filled equipment located on the jetty loading platform (e.g. oil-filled transformers, marine loading arm hydraulic units, etc.). Additionally, kerbs with scuppers are provided around the perimeter of the jetty loading platform where LNG or other continuous containment kerbs are not present to allow the drainage of water accumulated from precipitation. The jetty loading platform is designed for the combination of loads that will produce the highest forces for each structural component. Loads include dead loads, live loads, wind loads, wave loads, ice loads and seismic loads. Design load factors should be used pursuant to all applicable standards.

3.5 TRESTLE

The approach / pipe trestle connects the jetty loading platforms to the land and includes sufficient space to accommodate the pipe rack (including pipe loops), cable trays, the access road and buildings.

If the approach / pipe trestle is to be a pile supported structure the elevation of the trestle is established in a manner to allow the crests of the design waves to pass clearly under the structural elements of the approach / pipe trestle decks to prevent uplift loads from being imposed on the structure.

If the approach trestle is a rock causeway, it should be of sufficient height to prevent waves breaking over the pipes.

3.6 DOLPHINS IN GENERAL

The LNG jetties include pile supported breasting and mooring dolphins appropriately spaced on both sides of each respective jetty loading platform to support safe and efficient mooring of the full range of LNG carriers. Each dolphin is designed to deflect without causing movement in other nearby structure. The dolphins are supported by either large mono-piles or multiple raked piles. A total of four breasting dolphins are anticipated to be required for each jetty loading platform. The number of mooring dolphins required at the jetty loading platform is as follows:

- Six where only a small range of LNG carriers are to be moored
- Seven where a large range of LNG carriers can be moored with the bow always pointing out to sea.
- Eight where a large range of LNG carriers can be moored with either the bow or stern pointing out to sea.

Some general conditions are

- Breasting dolphin spacing, to locate on the parallel body side about 25% to 45% of Length Overall (LOA)
- Inner mooring dolphin spacing, 75 to 85 % of LOA
- Intermediate mooring dolphin spacing, 105 to 115% of LOA
- Outer mooring dolphin, 130 to 140% of LOA
- Fender stand-off distance, between 2 and 4m
- Quick release hook (QRH) elevation, about 1.3 to 1.8m above the dolphin deck
- QRH setback from dolphin edge, about 1.8 to 2.3m
- Mooring dolphin setback from the berthing line, 35 to 50m. This allows room between the mooring dolphins and the LNG carrier for tugs and line boats.

3.7 BREASTING DOLPHINS

The breasting dolphins are laid out in a manner such that two each are located on either side of the jetty loading platforms. They are optimally spaced to accommodate the full range of design LNG carriers to allow for appropriate arrangement of spring lines and to allow for full fender contact with the flat body of the range of design LNG carriers in the laden and ballast conditions. Moreover, the arrangement of the breasting dolphins should accommodate an optimum mooring line length of 35 to 50 metres from the LNG carrier's fairleads to the mooring hooks and the elevation of the mooring hooks should be such that the vertical angle from the LNG carrier's fairleads to the mooring hooks is no greater than 30 degrees from horizontal for the full range of design LNG carriers.

The breasting dolphins will each be outfitted with a suitable fender system to accommodate the berthing loads and wind loads on the LNG carriers as well as double quick release mooring hooks for making up spring lines from the LNG carriers. The capacity of each mooring hook is typically 150 metric tonnes. Additionally, each breasting dolphin is outfitted with an open backed accommodation ladder to allow access from line handling boats. At the top of the accommodation ladder there is a security gate.

A LNG carrier approach monitoring system with a suitable pilot display board is required. The pilot display board is located on the inner aft mooring dolphin at each berth and must be clearly visible from the LNG carrier's bridge wing during berthing manoeuvres. The fender system outfitted on each breasting dolphin consists of energy absorbing rubber fender elements, on steel frontal panels, and necessary mounting and restraining hardware. The mooring hook assemblies outfitted on each breasting dolphin will be comprised of a double quick release hook with an integrated capstan to aid in handling lines. The mooring loads are determined based on mooring analyses for the full range of design LNG carriers. The structures are designed to resist the resulting loads at the range of horizontal and vertical angles established for the position of each LNG carrier at the berth, and the location of the mooring equipment of each LNG carrier. The static mooring analyses should be performed utilising the typical wind and current design criteria and dynamic mooring analyses have been carried out to determine operational loading with the consideration of wind, current and waves. Typical wind conditions used in design are :

Wind: 25 knot velocity during berthing
Wind: 40 knots when LNG loading arms are connected
Wind: 60 knot velocity from all directions when moored

The breasting dolphins are designed for the combination of loads that will produce the highest forces for each structural component. Loads include breasting loads, mooring loads, dead loads, live loads, wind loads, wave loads, ice loads and seismic loads.

Suction effects from passing ships may alos need to be considered.

3.8 MOORING DOLPHINS

The mooring dolphins are spaced to accommodate the full range of design LNG carriers to allow for appropriate arrangement of breast lines. Moreover, the arrangement of the mooring dolphins should accommodate an optimum mooring line length of 35 to 50 metres from the LNG carrier's fairleads to the mooring hooks and the elevation of the mooring hooks should be such that the vertical angle from the LNG carrier's fairleads to the mooring hooks is no greater than 30 degrees from horizontal for the full range of design LNG carriers.

The mooring dolphins will each be outfitted with mooring hooks to connect mooring lines from the LNG carriers. The inner-most mooring dolphins will be outfitted with triple quick release mooring hooks and the other mooring dolphins will be outfitted with quadruple quick release mooring hooks. LNG carriers less than 170,000 cubic metres in capacity will utilise 16 mooring lines while all LNG carriers greater than 170,000 cubic metres in capacity will utilize 18 to 20 mooring lines. The capacity of each mooring hook is about 150 metric tonnes. Additionally, each mooring dolphin is outfitted with an open backed accommodation ladder to allow access from line handling boats. At the top of the accommodation ladder there is a security gate.

The mooring hook assemblies are outfitted with an integrated capstan to aid in handling lines. The mooring loads are determined based on mooring analyses for the full range of design LNG carriers. The structures are designed to resist the resulting loads at the range of horizontal and vertical angles established for the position of each LNG carrier at the berth, and the location of the mooring equipment of each LNG carrier. The analyses are typically perfomed for a 60 knot wind velocity having a 30-second duration coupled with the applicable current and wave forces.

The mooring dolphins are designed for the combination of loads that will produce the highest forces for each structural component. Loads include mooring loads, dead loads, live loads, wind loads, wave loads, ice loads and seismic loads.

3.9 CATWALKS

Catwalks are provided to connect all of the mooring and breasting dolphins to the respective jetty loading platform. The catwalks are provided with a clear access width of 1.2 metres and they are outfitted with heavy gauge handrails and anti-slip grating or appropriately finished concrete surfaces. Additionally catwalks are outfitted with a 600 mm wide extension outboard of the walking surface to support the installation of cable trays.

3.10 JETTY CONTROL BUILDING

A typical jetty control building has two floors with mechanical and electrical switchgear located on the bottom floor and the necessary jetty control instrumentation located on the upper floor. The upper floor will have panoramic views of the jetty loading platform(s), which will be protected with blast-proof glass or similar. The size of the building will allow a minimum clearance of 1.5 metres around the perimeter of all equipment for maintenance of equipment.

3.11 FIRE WATER PUMP SHELTERS

LNG jetties must be provided with adequate fire fighting facilities. The usual arrangement is for the jetty to be connected to an onshore freshwater firewater tank, with the connecting piping either in glass reinforced plastic or galvanised steel. If a major fire occurs onshore or on the jetty, the firewater system is topped up from seawater firepumps fitted on the jetty. The jetty fire pumps are normally diesel powered.

The jetty fire water pump is covered with a shelter to provide protection during maintenance. The fire water pump shelters typically feature a removable roof to allow for maintenance of the equipment they house. Each fire water pump shelter will accommodate a diesel powered fire water pump, inclusive of a diesel fuel tank. The shelter will be located over pipe caissons that extend vertically down from the platform towards the water below. One pipe will extend into the water and serve as a caisson for the pump to take suction from the sea, which will be piped to the fire water system and the second pipe will be used as an outlet for seawater during fire pump testing.

4. SAFETY

4.1 LNG CARRIER IN TRANSIT

During transit to and from the LNG terminal all other marine traffic has to be one nautical mile clear of the LNG carrier, fore and aft.

Safety reports are required to take account of external events which could lead to a major accident at the LNG site. An incident on a LNG carrier must be taken into account in the safety report to the extent that it could affect the safety of the LNG jetty, see reference 1.

Harbours used for the unloading or loading of LNG are managed by statutory harbour authorities. Harbour authorities control the marine traffic into and through the harbour and the berthing and moving of LNG carriers and are bound by the Port Marine Safety Code.

4.2 PIPELINES AND STORAGE

The safety reports should include an assessment of the integrity of the LNG pipelines within the LNG terminal and on the jetty. The design takes into account the low temperature service of the pipelines.

4.3 AIR GAP

An air gap of 1.5 metres is maintained between bottom of deck structures (i.e., loading platform deck, trestle deck) and the maximum wave crest elevation for the 100 year event.

The air gap requirement does not apply to the bottom of mooring and breasting dolphin decks. However, the dolphin top elevation must be above the highest waves. The dolphin top deck must have a non-slip finish.

4.4 IMPOUNDING BASIN

In case of a LNG spillage from the loading arms LNG needs to be channelled into an impounding basin locating just behind the loading platform. A typical volume would be about 150 cubic metres but this depends on the maximum spillage from a LNG loading/unloading arm before the safety valves close.

4.5 LIGHTNING PROTECTION MAST

Lightning protection is necessary on the loading platform to ensure there are no electrical sparks transmitted through the LNG loading/unloading arms. Two lightning protection masts would be required per loading platform. The upper elevation of the lightning mast is about twice the height of the LNG loading/unloading arms. The lightning mast requires a connection to earth to perform its protective function which is achieved by fitting a conductor through the concrete loading platform to be attached to a steel pile underneath.

5. DESIGN METHODOLOGY

5.1 MARINE EQUIPMENT

The following typical materials and equipment is purchased from specialised vendors:

- Marine loading arms
- Fender systems
- Quick release hooks and capstans
- Docking aid systems
- Mooring line load monitoring systems
- Tower and Column Gangways
- Elevated fire monitor towers
- Aids to navigation

5.2 JETTY TOPSIDE DESIGN

The jetty substructure supports equipment, piping, electrical and instrument cables. Typical piping on a LNG jetty trestle is shown in Table 4.

In order to prevent damage from piping surge loads, caused by the application emergency shut down valves, piping anchors are required

Because contraction occurs in the LNG pipes, at -167 degrees C, contraction loops are required about every 200 metres along the jetty. Piping guides are provided at frequent intervals. Piping anchors are typically provided mid-way between loops and at the middle of the loop.

5.3 ICE AND COLD REGION DESIGN

When developing LNG import/export facilities in locations where air temperatures are below zero for long periods the following items need to be considered:

- Metocean condition definition
  - Ice
  - Wind
  - Wave
  - Tides
  - Storm surge
  - Tsunamis
- Development of ice loads
  - Floe impact
  - Uplift
  - Ridging
- Dredging in ice conditions
- Navigation assessment
- Ice scour analysis of the seabed, in particular around the foundation piles
- Ice-induced abrasion evaluation on the pile foundations

There are few ice strengthened LNG carriers so tugs are required to break out any ice that is close to the LNG carrier or the LNG jetty.

5.4 DREDGING

Ideally a site should be chosen which requires no dredging, but where required the following are to be considered:

- Design
  - Maintenance dredging
  - Channel and berth deepening
  - Contaminated sediment remediation
  - Beneficial reuse of dredged material
  - Reclamation design
- Sediment assessment, modelling and management
  - Site characterization
  - Sediment/contaminant resuspension and transport
  - Sediment loss at the dredge
  - Chemical fate and effects
  - Remediation technologies
  - Geochemistry and toxicology
  - Dredged Material Management Plans
  - Risk assessment and management
  - Monitoring and regulatory agency coordination
- Dredging engineering
  - Plans and specifications
  - Cost estimating and cost management
  - Contractor selection
  - Contract management
- Environmental Support
  - Permitting alternatives and analysis
  - Environmental documentation
  - Permit acquisition
- Beneficial reuse
  - Habitat creation and restoration
  - Land creation
  - Beach restoration

5.5 BREAKWATER

Ideally a site should be chosen which requires no breakwater. If a breakwater is required for the Material Offloading Facility (MOF) or the LNG jetty then this will have a significant impact on the construction schedule and costs.

5.6 LNG CARRIER MANOEUVRING

The safe navigation of the LNG carriers is developed in conjunction with the local Port Authority and this forms part of the planning permission process for the LNG plant. In addition, as there is a safe zone around the LNG carrier whilst underway and whilst moored at the LNG jetty, other users of the port also need to be consulted.

Navigation studies cover approach and departure from the LNG jetty. The safe navigation starts at the location where the marine pilots board the LNG carrier during the approach to the LNG jetty and finish where the marine pilots leave the LNG carrier after departure. Areas to be considered are:

- Approach Channel analysis and design
- Berthing and turning basin layout
- Dredging studies

LNG carrier manoeuvring is studies through simulation:

- Statistical and flow modelling of LNG carrier traffic
- LNG carrier traffic analysis
- Fast-time simulation
- Real-time manoeuvring simulation
- Full mission bridge simulation management

Navigation aids are fitted to the monitor and record the following:

- Weather monitoring
- Pilot display board. There are two pilot display boards per LNG jetty, usually fitted on a mooring dolphin and they show LNG carrier approach speed and LNG carrier angle relative to the berth
- Fixed navigation aid design
- Permitting

5.7 MOORING AND BERTHING

Mooring analysis includes the following:

- Jetty terminal mooring design
- Static mooring analysis
- Frequency and time-domain dynamic analysis
- Downtime analysis
- Fatigue analysis
- Mooring arrangement optimisation
- Passing ship analysis
- LNG carrier manifold motions at berth
- Weather operating limit determination
- Mooring hardware selection

Berthing analysis includes the following:
- Geometric evaluations of different size LNG carriers
- Fast-time manoeuvring simulation
- Real-time manoeuvring simulation

The Fender system design includes:
- Berthing impact energy analysis
- Fender type selection and design
- Structural system component design
- Elevation of the fenders

5.8 LNG CARRIER SHIPPING AND LNG STORING SIMULATION

LNG carrier logistics and storage (import and export LNG terminals) analysis is used to determine the number of LNG berths. The berth optimisation studies include:
- Required LNG storage tank capacity
- Required LNG carrier fleet size
- Wait time for LNG carriers
- Weather downtime of LNG carriers
- Berth occupancy

The LNG Berth availability evaluation includes:
- Met ocean conditions
- Tug availability
- Marine traffic
- Pilot availability and delays to pilot boarding

The traffic studies include:
- In-bound and out-bound movements
- Various LNG carrier types
- Port regulations
- Pilot assistance
- Restrictions on two-way traffic
- Wind, tide, visibility, current, and/or night transit restrictions

5.9 GUIDANCE ON DESIGN

Each country has its own guidance on jetty design and a few of these are as follow:
- UK – Maritime Works - BS6349
- Australia – Guidelines for the design of maritime structures –AS 4997
- USA – Coastal engineering manual – EM1110 (US Corps of Engineers)
- Germany - Recommendations of the Committee for Waterfront Structures Harbours and Waterways – EAU2004

In addition there are technical reports from The World Association for Waterborne Transport Infrastructure (PIANC.

Further guidance is provided by the Society of Gas Tanker operators and terminal operators (SIGGTO). In addition the Oil Companies Internation Marine Forum (OCIMF) published LNG jetty guidelines.

6. INTERFACE INFORMATION

6.1 LNG CARRIERS

The following information is required from LNG carriers for LNG berth design:
- Length between perpendiculars
- Breadth
- Depth
- Centre manifold location relative to amidships
- Height of manifold above LNG carrier keel
- Minimum draft and displacement
- Maximum draft when loaded with LNG and associated displacement
- Wind area at maximum draft
- Wind area at minimum draft

The principal dimensions of typical LNG carriers are shown in:
- Table 1 shows Large LNG carriers
- Table 2 shows Medium Membrane containment system LNG carriers
- Table 3 shows Medium Spherical containment system LNG carriers

6.2 LNG JETTY

The following information is required for the LNG berth design for the LNG carrier:
- Water depth extreme low tide at the berth, in the turning circle and in the navigation channel
- Water depth at extreme high tide at the berth
- Minimum and maximum height of LNG loading/unloading arms
- Allowable lateral movement of the LNG loading/unloading arms
- IMO fire water connection
- Pilot display board locations on the jetty
- LNG carrier to shore connection details
- Location of the tower gangway
- Available tug details
- Location and elevation of mooring dolphins
- Location and elevation of breasting dolphins
- Quick release hook capacities
- Maximum and minimum elevation of fenders

6.3 LNG CARRIER SIZE TRENDS

LNG terminals have been governed by large scale export and import of LNG, and hence the facilities have been developed in conjunction with large LNG carriers. Many of the world's LNG terminal facilities are limited to bigger LNG Vessels in the range 87,000 cum to 265,000 cum. However there are the following trends which need to be considered in future LNG terminal design
- the trading of smaller volumes of LNG
- LNG fuelled ships

There are several LNG carriers in the range 10,000 cum to 30, 000 cum, which if used at a large scale LNG terminal will require modification to existing breasting dolphins. However the manifolds would need to be at a very high elevation on the small LNG carrier to match the manifold heights on large LNG carriers. Loading of LNG fuelled ships is expected to be cyrogenic hoses.

7. CONSTRUCTION LOGISTICS

Greenfield LNG marine terminal facilities are often developed in remote locations, inaccessible by roads and far from large metropolitan areas. Initially a pioneer dock must be set up in the first few months of the construction, and is used to import equipment and accommodation for site preparation.

If large pieces of equipment are to be delivered to the LNG terminal construction site by sea then a Material Offloading Facility (MOF) is required which can berth large geared vessels, i.e. with cranes, Heavy Transport Vessels.

Design requirements include:

- Site evaluation and selection
- Construction harbour planning
- Pioneer Dock planning and design
- Crew boat and tug berthing facilities in operation
- Dock facilities for Heavy Transport Vessels and barges
- Transport and construction logistics

Large prefabricated piping modules are often the preferred method of construction for LNG facilities in remote areas. Therefore, most equipment and materials for these projects has to be brought in using marine transport. The Material Offloading Facility (MOF) requires structures that are designed for very heavy loads and can accommodate specialized Heavy Transport Vessels (HTV) and self-propelled modular transporters (SPMTs), so berths are needed at the MOF for flat deck Heavy Transport Vessels

8. CONCLUSIONS

The design and construction of a LNG jetty is a crucial part of the LNG plant construction schedule.

The safe operation of the loading arms is critical to the safety of the LNG facility during operations. The loading platforms are designed for wave, wind, tidal current, seismic and LNG piping loads and potential LNG spillage.

The design range of LNG carriers needs to be defined at an early stage in the design so that the breasting and mooring dolphins can be correctly located.

9. ACKNOWLEDGEMENTS

The author thanks his colleagues in CB&I for their assistance in preparing this paper. However the paper represents the opinion and views of the author, and does not necessarily represent those of CB&I, or any of its subsidiaries.

10. REFERENCES

1. UK HSE rules on consent and operational issues for LNG terminals

11. AUTHORS BIOGRAPHY

Alan Crowle holds the current position of principal naval architect at CB&I, London. He is responsible for naval architecture activities for offshore projects and the coordination of LNG jetty design. His previous experience includes steel jackets, semisubmersibles, FPSOs, large offshore and onshore modules, and near-shore facilities.

Table 1 Typical Large LNG Carriers

PARAMETER	TYPICAL	QFLEX	QMAX	TYPICAL	LARGE
Containment	Membrane	Membrane	Membrane	Spherical	Spherical
Capacity (m^3)	147,640	210,100	266,000	145,400	177,000
Displacement (Laden) (t)	104,000	143,000	177,000	105,000	128,000
Displacement (Ballast) (t)	78,000	112,000	142,000	76,000	94,000
LOA (m)	285.0	315.0	345.0	289.5	300.0
LBP (m)	274.0	303.0	332.0	277.0	286.5
Beam (m)	43.4	50.0	53.8	49.0	52.0
Moulded Depth (m)	26.0	27.0	27.2	27.0	28.0
Laden Draft (m)	11.4	12.0	12.0	11.4	12.5
Ballast Draft (m)	9.6	9.6	9.8	9.4	11.5
Max Freeboard (m)	16.4	17.4	17.4	17.6	16.5
Manifold Offset from Midship (m) (+ve fwd)	3.3	3.5	3.5	3.4	3.4
deck(m)	5.0	5.0	5.0	5.5	5.5
Manifold Offset from Midship (m) (+ve fwd)	-1.0	4.0	5.0	15.6	12.8
Manifold Centreline to Flat Hull Limit Forward (m)	53.2	68.5	76.7	41.3	74.2
Manifold Centreline to Flat Hull Limit Aft (m)	77.6	72.5	83.8	78.1	117.3
Min Breaking Load of Mooring Lines	125	135	137	125	130
Number of Mooring Lines	16	18	20	16	18
Windage Area laden (side) (m^2)	6,785	7,130	8,759	7,680	8,686
Windage Area laden (end on) (m^2)	1,445	1,510	1,612	1,530	2,029
Windage Area ballast (side) (m^2)	7,399	7,776	9,552	7,835	9,472
Windage Area ballast (end on) (m^2)	1,576	1,647	1,741	1,654	2,213

Table 2 Typical Medium Size Membrane containment system LNG Carriers

Containment	Membrane	Membrane	Membrane	Membrane	Membrane
Capacity (m^3)	65,000	77,679	125,000	130,000	132,500
Displacement (Laden) (t)	52,900	60,840	103,070	99,350	99,880
Displacement (Ballast) (t)	47,119	51,315	95,774	84,899	90,545
LOA (m)	216.1	248.0	278.8	268.5	286.9
LBP (m)	205.0	230.0	262.6	257.0	275.0
Beam (m)	34.0	34.8	41.0	43.0	41.8
Moulded Depth (m) to upper deck at midships	21.3	21.2	25.8	27.0	28.1
Laden Draft (m)	9.2	9.9	11.2	11.0	10.7
Ballast draft (m)	8.2	8.4	10.4	9.4	9.7
Maximum Freeboard (m)	13.1	12.9	15.4	17.6	18.4
Manifold from LNG Carrier side shell (m)	3.0	2.9	3.5	3.6	3.5
Manifold above upper deck(m)	4.2	6.6	7.0	4.1	3.0
Manifold Offset from Midship (m) (+ve fwd)	2.9	-9.8	1.1	5.5	3.4
Manifold Centreline to Flat Hull Limit Forward (m)	44.0	54.6	57.0	44.6	72.0
Manifold Centreline to Flat Hull Limit Aft (m)	60.0	23.4	85.0	75.1	73.4
Min Breaking Load of Mooring Lines (t)	90	100	110	110	110
Number of Mooring Lines	10	12	12	12	12
Windage Area laden (side) (m^2)	945	616	1,655	1,357	1,116
Windage Area laden (end on) (m^2)	3,529	2,994	5,863	5,512	5,924
Windage Area ballast (side) (m^2)	1,061	730	1,781	1,588	1,231
Windage Area ballast (end on) (m^2)	3,962	3,550	6,310	6,450	6,535

Table 3 Typical Medium Spherical containment system LNG Carriers

Containment	Spherical	Spherical	Spherical	Spherical	Spherical
Capacity (m^3)	87,000	125,000	125,000	125,000	137,000
Displacement (Laden) (t)	74,380	92,160	93,240	93,240	106,000
Displacement (Ballast) (t)	66,661	81,742	76,987	77,487	87,523
LOA (m)	249.5	283.0	272.0	272.0	293.0
LBP (m)	237.0	269.0	259.0	259.0	280.0
Beam (m)	40.0	44.6	47.2	47.2	45.8
Moulded Depth (m) to upper deck at midships	23.0	25.0	26.5	26.5	25.6
Laden Draft (m)	10.6	11.5	10.9	11.0	10.9
Ballast draft (m)	9.5	10.2	9.0	9.1	9.0
Maximum Freeboard (m)	13.5	14.8	17.5	17.4	16.6
Manifold from LNG Carrier side shell (m)	4.5	3.5	3.5	3.5	2.8
deck(m)	5.9	5.3	4.0	4.0	3.2
Manifold Offset from Midship (m) (+ve fwd)	-11.8	30.7	10.5	-10.5	-3.5
Manifold Centreline to Flat Hull Limit Forward (m)	58.5	64.6	45.3	35.5	67.3
Manifold Centreline to Flat Hull Limit Aft (m)	33.6	87.5	68.2	80.2	64.1
Min Breaking Load of Mooring Lines (t)	100	110	110	110	115
Number of Mooring Lines	10	12	12	12	14
Windage Area laden (side) (m^2)	1,037	1,357	1,433	1,488	1,286
Windage Area laden (end on) (m^2)	4,741	6,949	6,397	6,441	6,707
Windage Area ballast (side) (m^2)	1,157	1,530	1,735	1,790	1,558
Windage Area ballast (end on) (m^2)	5,290	7,835	7,747	7,750	8,123

Table 4 Typical piping and cable racks on a LNG trestle

TYPE	NUMBER OF LINES	DIA (NPS)	DIA (NPS)	MATERIAL	INSULATION
		inches	mm		mm
LNG LOADING LINE	1	36	DN900	Stainless steel	280
LNG VAPOUR RETURN	1	24	DN600	Stainless steel	200
LNG RECIRCULATION	1	12	DN300	Stainless steel	180
SERVICE LINES for air for tools, instrument air, drinking water, nitrogen, cooling water for instruments	5	4	DN100	Carbon steel	0
FIRE WATER LINES	2	16	DN400	GRP	0
900mm WIDE CABLE RACKS	8			stainless steel	

Figure 1 216,000 cum LNG Carriers at 2 berth Terminal Showing

- 2 loading platforms
- Access trestle with pipes and roadway
- Piping loops

Figure 2 Typical LNG berth Showing

- Breasting dolphins and fenders
- Mooring dolphins
- Loading platform with LNG unloading arms, gangways, fire monitor towers and lightning protection masts

SMALL SCALE LNG BUNKERING SOLUTIONS

A Bjoern, Kosan Crisplant Denmark

SUMMARY

The island of Samsø reached an important milestone on its way to a greener future when its new ferry was bunkered with LNG for the very first time.

Kosan Crisplant small scale LNG bunkering solutions for ferries, short sea ships and inlands vessels is evolving into new versions and variations as the small scales market develops.

1. INTRODUCTION

The zero-emission LNG bunkering facility makes it possible to fuel a ferry for a whole day's operation in less than 30 minutes, all handling activities included. The bunkering facility is fully automatic and requires no permanent staff on shore. With its modular design, the LNG bunkering facility can be dismantled and moved to another location in less than a week.

2. THE BUNKERING SYSTEM

The bunkering system is built in modules. Everything are made prefab and tested in the work shop. After the test and 3rd part approval, the modules are shipped out for fast site erection.

Figure 1: Pumping container under construction in the workshop

Figure 2: Ramp under construction

2.1 SIMPLE AND SAFE OPERATION

Approx. 15 minutes before the ferry arrives to the facility by the quay, a signal is sent from the ferry to the control system of the bunkering plant. This activates the system and the LNG road tanker controls to start the pre-cooling sequence of the transfer pump.

When the ferry has docked, a ferry crew member connects the EMS (emergency shut-down) cable, takes the loading hose from its parking position and connects it to the ferry bunkering coupling, and starts the bunkering process on the HMI control panel. After the desired amount of LNG has been transferred, the process is stopped on the HMI control panel, thereby stopping the pump and closing the valve on the road tanker liquid outlet line. When 'bunkering valve close' is activated on the HMI control panel, the liquid is led back to the road tanker, thereby lowering the pressure in the bunkering line. When the pressure in the bunkering line has reached a pre-set value, the HMI will display 'disconnect bunker hose'. The loading hose is then disconnected and placed on the parking connection again.

As the ferry leaves, the remaining liquid in the lines will be vaporized and displaced to the road tanker through the vapour line due to thermal expansion and lower pressure in the road tanker. Finally, all valves are closed, putting the system 'to sleep' under vapour pressure. During 'sleep', the pressure in the system is kept on a constant level, thereby preventing any emission to the atmosphere. Kosan Crisplant's complete turnkey solution includes the following main items:

- Cryogenic transfer pump unit built into a 20 feet container
- Piping system incl. specially designed LNG dry couplings
- 2 specially designed LNG road tankers
- Parking ramp for road tanker
- Control system incl. safety system
- HMI control panels
- Power installation
- Theft-proof electrical panel placed outside Eex zone 1
- Compressed air installation placed outside Eex zone 1
- Fire and gas alarm system
- Fire fighting equipment
- Fencing, outdoor lighting and lighting in container

2.2 TECHNICAL FACTS

Max. bunkering time:.......................................30 minutes

Pumping time during bunkering process:........15 minutes

Container size:...20 feet

Fuel consumption (ferry):..........Min. 10 m^3 LNG per day

Pump capacity:................60 m^3/hour (1000 litres/minute)

Differential pressure range:.................................4-6 bar

Volume of LNG road tanker (geometric/operational):.............................56 m^3/44,5 m^3

Necessary connections:.........3 x 400 V power supply and compressed air installation

Power consumption:..20 kW

Patented solution:..............Danish Patent No. PR 178151

Figure 3: Assembly of modules on site

3. THE LOGISTIC SOLUTION

Special designed cryogenic tanker, which meets the weight requirements for motorway driving in Germany, is transporting LNG from Rotterdam. The solution involves two alternating LNG tanks on the quayside in Hou. Every three days, the empty tank is replaced with a full one from Rotterdam. The tanker parked on the quayside is considered as a permanent tank installation, with all the requirements that follows, on the road the tanker is an ADR vehicle.

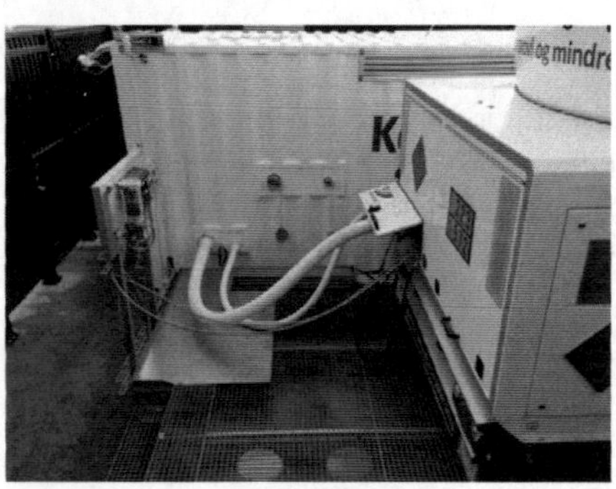

Figure 4: Road tanker connected to the pumping station

4. NON EMISSION - A CHALLENGE

The pump and connecting pipe installation are built into a 20 feet container. Connecting hoses are all equipped with Mann Tek LNG dry couplings, which have no product release during connection/dis-connection. The pump is sized so the total bunkering period is 15 min, including the handling. The actual pumping time will be 10 min, equals 60 m3/h (1000 Litres/min). After the bunkering the liquid in the onshore pipe system will be displaced to the road tanker, all connections will be closed and the plant will be under vapour pressure. Volume in the piping system is around 100 Litres.

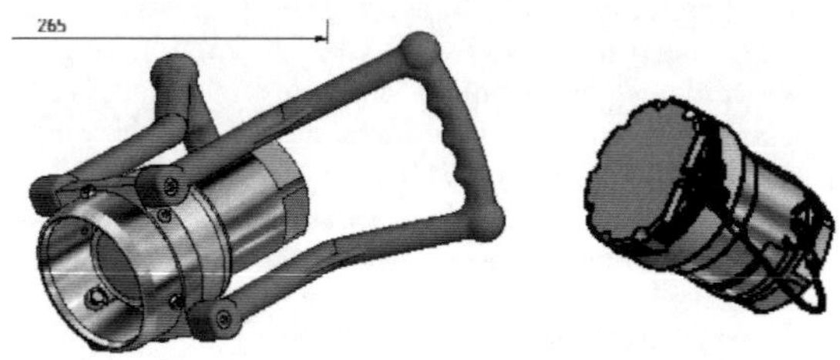

Figure 5: Mann Tek dry couplings

A Thermodynamics calculation was done to proof that the tanker logistic and bunkering operation was possible. The heat impact to the tanker is coming from pump operations, vapour recovery and the quality of the super insulation of the tanker.

The results of the calculation are a description of pressure and temperature in the tanker during the departure from Gate Terminal to the Ferry and return. This is shown in the two graphs below.

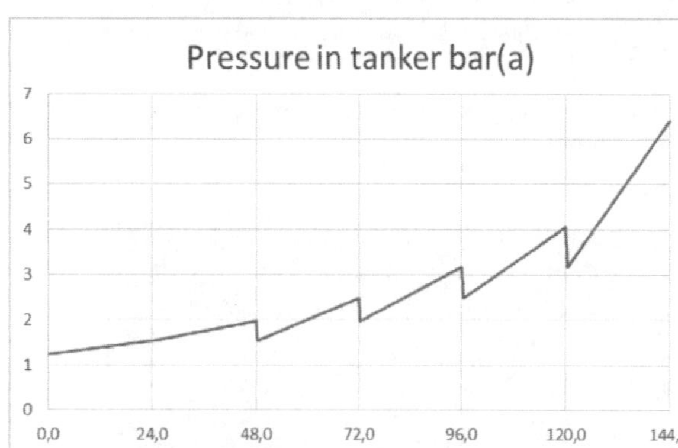

Figure 6: x axis shows the amount of hours from loading at Gate Terminal

The graph shows that the absolute pressure will not exceed 6.t 4 bar(a), provided that a small amount of LNG(liquid) is left in the tank.

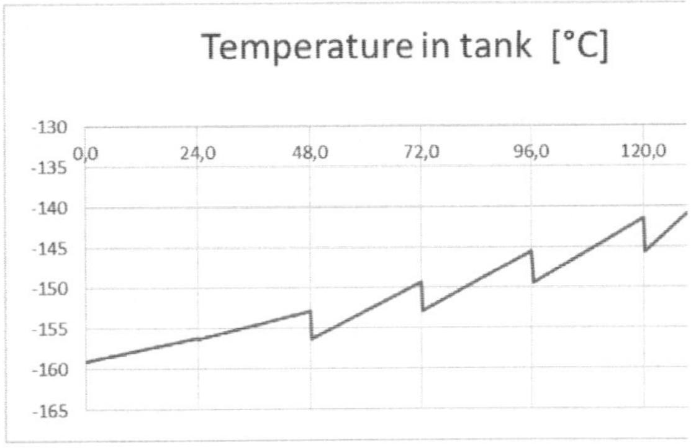

Figure 7: x axis shows the hours from loading at Gate Terminal

The temperature in the tank follows the equilibrium between vapour and liquid. It is shown that the temperature is not exceeding - 135°C. If the tank trailer was empty the raise of the temperature was more rapidly.

5. AUTOMATION

The bunkering system is the ever first fully automatic one man operated.

Figure 8: Control panel being tested

The process is controlled by a central PLC, in which the controls is made "fail safe" in three levels.

Level 1: Power supply from an on-line UPS (runtime approx.. 30 min)

Level 2: The PLC control is builded "Fail Safe" so if an unintended incident happens and the PLC loses the connection/control, the all outputs will be deactivated and the LNG valves will, as default, be closed. The complete system will still be "safe" as trapped LNG can be evacuated through safety valves to a blow-off mast.

Level 3: A GSM module with 8-9 channels for differential SMS alarm to service centre.

6. OPERATION

For both the tanker and ship connection/disconnection a special sequence programme has been developed. The operator shall follow this sequence precisely every time. If this sequence is not followed the system will enter into "fail safe" mode and the operator must "reset" the system and start all over again. This very conservative way of operation ensures that the system will always be working correct and accidents are avoided

Figure 9: Control lamp for operational assistance

Both the tanker and the ferry are connected to the control system and ESD system. The connection to the ferry consists of two cables with pull-away sockets. One is connected to the Ferry´s MOD bus system and will send signal for opening and closing of the bunker valve. The other connection is hard wired to the ferry´s ESD system and connects to the land based system.

Furthermore Fire and Gas detection is installed in strategic places to detect gas leaks and potential fires.
The F/G alarm central is connected to the local fire department where an alarm signal is coming by gas leaks and fire

7. CONCLUSION

The bunkering system for the Samsoe Ferry has now been in operation since May 2015. Bunkering is performed every day. Approximately 10 m3 is bunkered each day in order to reduce weight. The first operation manual from the ship yard required a staff of 3 persons to do the bunkering. Now, 1 person ca do this alone and safe. There have not been any considerable technical problems, during this period of operation. 4 inspection visits are made per year, for checking and inspection of the installation.

Kosan Crisplant has continued the development of the LNG solutions for ship bunkering. A system for connection of more tankers to one pump station and bunkering of higher volumes are under development.

Also standard tankers and ISO containers can be used in this solution. This will lower the CAPEX and be more flexible.

The marked for the small scale bunkering facilities are growing as nobody today want to invest in fixed assets host big permanent installations.

8. AUTHORS BIOGRAPHY

Anders Bjoern holds the current position of Project Director/COO at Kosan Crisplant. Is responsible for Project design office and execution.

LNG SHIP MK III CARGO CONTAINMENT SYSTEMS - TESTING AND INTERPRETATION OF RESULTS, IN-SERVICE, FOR SECONDARY BARRIERS

J R Maguire, Independent Consultant, UK

SUMMARY

There are many marine projects which involve the transportation, storage and processing of liquefied natural gas (LNG) in cargo containment systems (CCS). There are a number of different types of CCS, including membrane, MOSS and SPB. With respect to the membrane type, the predominant systems are designated NO96, Mk III and CS1, general descriptions being available from the designer, Gas Transport and Technigaz (GTT) from their web site www.gtt.fr. The particular CCS of interest here is the Mk III type, and the particular area of interest is the secondary barrier. To assess the in-service condition there are a number of different test types available, such as:

- pre-docking/at sea: Low Differential Pressure Test, and Thermographic Assessment of Membrane Integrity;
- during docking/alongside: Secondary Barrier Tightness Test, and global and local Acoustic Emission.

These different but complimentary test types will now be reviewed, with particular consideration given to interpretation of the test results.

1. INTRODUCTION TO THE CCS

The membrane tank is non-self-supporting and consists of a very thin layer or membrane (primary barrier) that derives its support from the adjacent hull structure via the interconnecting insulation material. The primary barrier is in contact with the cargo. Membrane containment systems must always be provided with a secondary barrier to ensure the integrity of the total system in the event of primary barrier leakage. The secondary barriers on Mk III ships **[Ref. 1]** are constructed from a composite material – triplex – which is, by nature, liquid tight but not necessarily 100% gas tight.

The space between the primary barrier and the secondary barrier is known as the inter barrier space (IBS) or the primary space. The space between the secondary barrier and the hull structure is called the insulation space (IS) or secondary space, see **Figure 1** below.

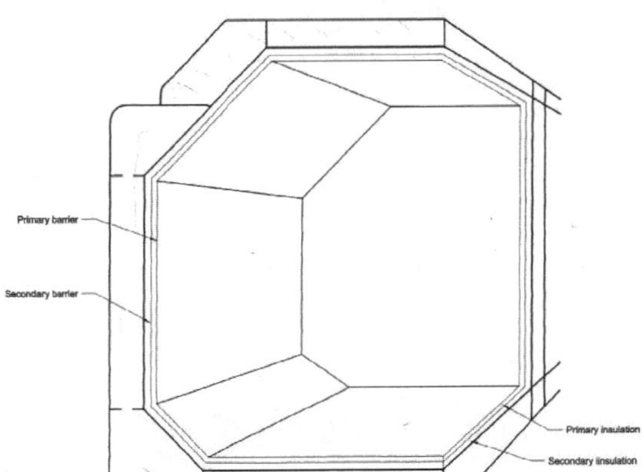

Figure 1: Cargo tank arrangement

Ship Classification Society Rules require the secondary barrier of the Mk III LNG Cargo Containment System to be periodically checked for its effectiveness using an appropriate method. There are two methods, currently recognised by all major Class Societies involved with LNG ships, for doing this, :

(a) to carry out a Secondary Barrier Tightness Test (SBTT);

(b) to carry out a Thermographic Assessment of Membrane Integrity test, combined with an Acoustic Emission (AE).

The designers of the Mk III system, Gas Transport and Technigaz (GTT), publish documents which describe SBTT **[Ref. 2]**, Thermographic Assessment **[Ref. 3]** and AE testing **[Ref. 4]**, as well as other forms of test, for example the Low Differential Pressure Test (LDPT) **[Ref. 5]** which will be described later. Recognised testing companies have to produce their own procedures (based on GTT procedures) which must be reviewed and approved by GTT. For example, Cryovision can undertake a TAMI© test **[Ref. 6]** which is their adaptation of the Thermographic Assessment of Membrane Integrity test.

2. TESTING OPTIONS & INTERRELATION

Ideally, there would be a single test which could check and simultaneously **detect, locate and quantify** any defects in the secondary barrier, but unfortunately the current state of marine technology does not permit this, instead one has to select more than one test for a comprehensive understanding. For example, SBTT can **detect and quantify**, but cannot locate defects, whereas AE can **locate** but cannot accurately quantify the magnitude (or contribution) of individual defects to the overall result.

Generally, tests fall into two categories, firstly those that can be carried out at sea (in service) **prior** to dockings, and secondly those that are carried out **during** dockings.

An advantage of the former – such as LDPT and/or TAMI at sea - is that they can give a valuable "heads up" preview of what is likely to be needed at docking, in terms of possible extent of repair needed and associated time and material quantities. However, as regards the "official" qualification test, as required by Class Societies, the currently recognised qualification tests are SBTT, and Thermal Assessment combined with AE in areas where Thermal Assessment is not reliable, namely in the very uppermost regions of the cargo tanks, such as the liquid and gas dome areas (see **Figure 2** below).

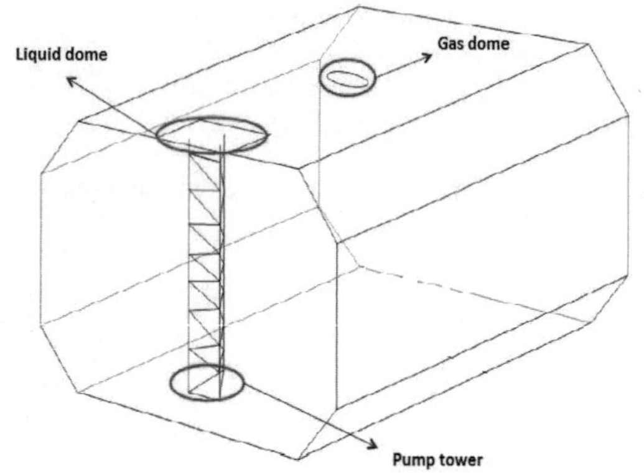

Figure 2: Example locations of liquid and gas domes

The liquid dome is the area where the pump tower crosses the decks. The gas dome, also at deck level, allows:

- for gas exchanges between the terminal and the tank during loading and unloading operations;
- to release gas to the fuel gas system.

The type of defects that are being looked for are small areas of localised glue debonding, as illustrated in **Figure 3 below.**

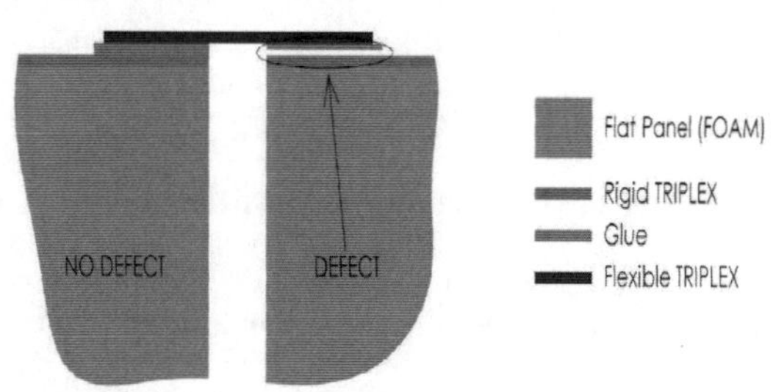

Figure 3: Localised debonding representation

3. LOW DIFFERENTIAL PRESSURE TEST

The SBTT test involves the application of a relatively "high" differential pressure across the secondary barrier through the application of a partial vacuum to the insulation space. SBTT can be carried out when the vessel is gas free and alongside, or in dry-dock, or if the vessel is at anchorage within protected waters.

The TAMI test involves the use of a relatively low differential pressure and the thermal gradient of the insulation on a loaded vessel to detect by thermal imaging the impact on the inner hull of the gas passing through the secondary barrier, and will be considered in more detail later in this paper.

To avoid (potentially damaging) high differential pressures, Samsung Heavy Industries (SHI) and GazTransport & Technigaz (GTT) have collaborated in developing an alternative approach to monitoring the effectiveness of the secondary barrier. This procedure, called Low Differential Pressure Testing (LDPT), can be carried out while the vessel is operating and therefore allows more frequent monitoring of the barrier membrane tightness. LDPT consists of two tests, both of which involve the application of relatively "low" differential pressures across the secondary barrier:

- LDPT-1 checks for secondary barrier porosity from the IS to the IBS (performed in unladen or laden conditions);
- LDPT-2 checks for secondary barrier porosity from the IBS to the IS (performed in unladen condition).

External leaks (not due to the integrity of the membranes but due to other outfitting) can have impact on both LDPT-1 & 2; in order to detect these leaks a pre-test must be performed by pressurising both IBS and IS at the same pressure and monitoring. Diagrams illustrating LDPT-1 and LDPT-2 are shown in **Figures 4 and 5** below.

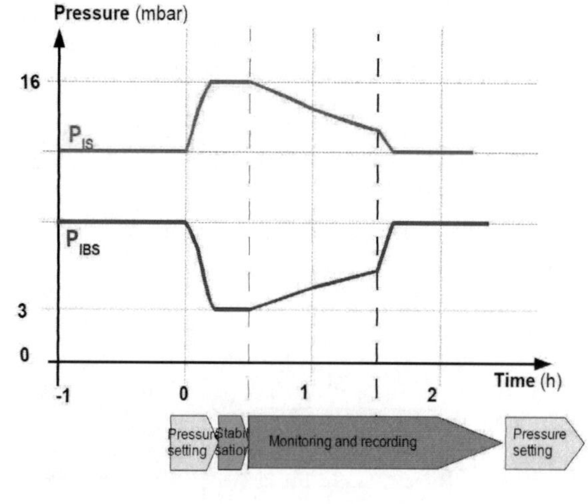

Figure 4: LDPT-1 Pressure vs. time

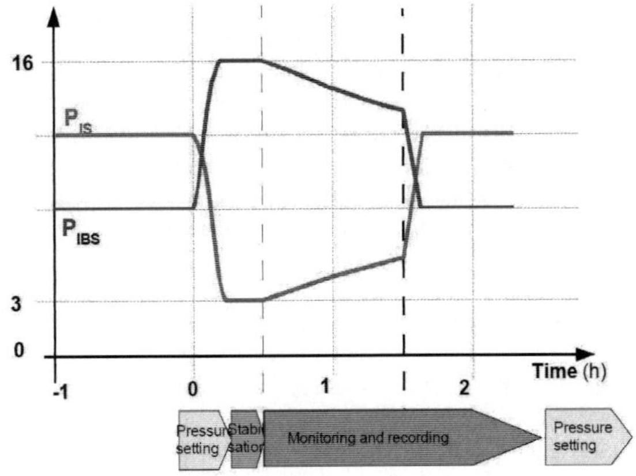

Figure 5: LDPT-2 Pressure vs. time

In recent years the LDPT test seems to have fallen out of favour as a qualifying testing method, but is still in use to monitor the evaluation of known leaks. As a result, LDPT will not be discussed further in this paper and instead more emphasis will be given to other, more popular, approaches.

4. SECONDARY BARRIER TIGHTNESS TEST

The Secondary Barrier Tightness Test (SBTT) is a critical part of the classification approval process and the test protocol is often agreed in advance of the docking, to avoid delays and adverse confrontations during the docking. The SBTT can be carried out when the cargo tanks are gas free and the vessel is alongside, or in dry-dock, or at anchorage in protected waters. Such testing normally forms part of the ship's Special Survey (every 5 years, on average).

The SBTT is carried out by establishing a relatively "high" differential pressure, of the order of ~500 millibars (mb), across the secondary barrier. The test supplies the data, which gives information about the global porosity of the secondary barrier of each cargo tank. The data obtained from the first test before delivery will be used as the benchmark for comparison with subsequent test results during the ship's life. In order to obtain test results that can be compared to the benchmark values the test conditions must be standardised, matching the conditions under which the benchmark data was obtained.

The SBTT commences with the inter-barrier space (IBS) open to atmosphere and the insulation space (IS) isolated and stabilized at a partial vacuum. The IBS is maintained at atmospheric pressure, while keeping the IS isolated and free to increase towards the IBS pressure (**Figure 6**). Any porosity of the secondary barrier would be from the IBS to the IS (this is important because if the differential pressure were reversed it could result in damage to the secondary barrier bonding).

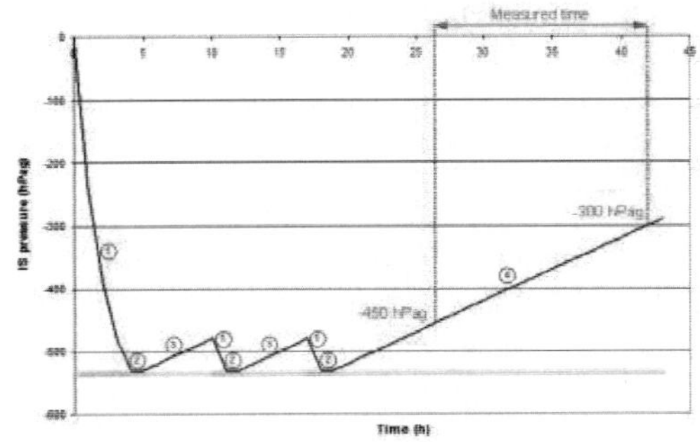

Figure 6: SBTT – IS pressure vs. time

The IS pressure is then monitored to determine the vacuum decay rate (VDR). At new build, typical VDR values would be expected to be not exceeding (and normally much lower than) 17 mbar/hr. The VDR is then used to agree what further action is required. Depending on what VDR is yielded by the SBTT further action may include all or some of the following:

- a controlled hearing test [**Ref. 4**];
- global acoustic emission (AE) mapping [**Ref. 4**];
- localised AE mapping (described below);
- TAMI, which can be performed at quayside with liquid nitrogen (LN2) as a cooling medium.

5. CUMULATIVE THERMAL CYCLES (CTC)

The VDR for any tank will vary with the "age" of the tank, age being defined in terms of fatigue cycles, rather than in calendar terms. The Cumulative Thermal Cycle factor (CTC) is the characterisation of the thermal cycling of the secondary barrier throughout the vessel's life. From "unloaded" to "loaded" to "unloaded", representing a typical "there and back" voyage, a thermal cycle of +20 deg C to – 163 deg C to +20 deg C might be experienced by each tank, equating to a CTC value of 1.0 for that particular round trip. The secondary barrier tightness condition of a tank is accepted if the VDR is less than a threshold value, as defined in a formula, and shown graphically in **Figure 7** below.

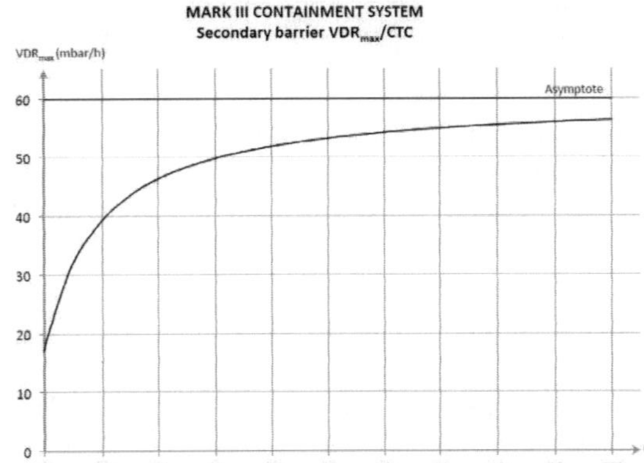

Figure 7: Relation between acceptable VDR and CTC

6. POST DOCKING VDR

Post docking, if the actual VDR is found to be greater than an agreed figure, when the vessel is afloat, LDPT should be carried out for reference purposes. This would be:

- in preparation for subsequent in-service secondary barrier monitoring;
- to check the IBS and IS could be maintained at the design differential pressure and therefore providing effective independent leak detection systems.

Vacuum Decay Rate (VDR) is affected by both normal and abnormal porosity of the secondary barrier (SB). Normal porosity of the SB contributes to the overall VDR but will not lead to liquid entering the insulation space (IS) in case of flooding of the IBS. Normal porosity cannot be detected through hearing (audible) tests or acoustic emission (AE) measurements. On the other hand, abnormal porosity (such as indications resulting from glue debonding) can be detected through hearing tests and AE measurements.

7. GLOBAL ACOUSTIC EMISSION

When VDRs yielded by SBTT indicate that further investigations should be undertaken a hearing test is first carried out. The hearing test should be carried out when the ship is otherwise "quiet", and ideally over a period of time to encompass a range of differential pressures. Global AE measurements are then undertaken in order to confirm the location of the hearing indications and detect any other indications that have not been detected by the hearing test. (Hearing tests are often effective when relatively "large" abnormalities are present, and acoustic emission tests can be effective when relatively "small" abnormalities are present.) The difference between "Global" and "Local" AE lies in the extent of coverage – "Global" refers to 100% of the tank wall area, whereas "Local" refers to limited areas, typically less than 10% of the tank area, where defects are considered to be more likely to occur.

A secondary barrier membrane with abnormal porosity will produce acoustic emissions when subjected to a differential pressure. The principal of AE measurement is to locate the position of the defect by measuring and mapping the AEs. To do this, global AE (entire tank) measurements are carried out by first installing low frequency acoustic probes (typically covering a broadband range, often of the order of 30 to 50 kHz) inside the cargo tank and/or outside in way of the double hull (ballast, cofferdam and trunk deck spaces). The probes are installed in a matrix pattern. The background noise is measured for calibration purposes and then a differential pressure is applied across the secondary barrier. The procedure is similar to that used for SBTT in that the IBS is maintained at atmospheric pressure and the IS isolated and stabilized at a partial vacuum. The IBS is maintained at atmospheric pressure, while keeping the IS isolated and free to increase towards the IBS pressure. Recording of the global AE measurements is carried out while the IS pressure is rising towards the IBS pressure (as shown in **Figure 8** below). Other sources of noise should be kept to a minimum during this phase. For confirmation of an indication, or more precise mapping of an indication, the spacing between the probes may be reduced.

Based on experience from vessels of the same class using established global AE procedures, an extended Special Survey docking might be required, if repairs are found to be necessary. The recent introduction of localised AE measurements can contribute significantly in reducing such periods, as discussed below. A time saving during docking is clearly of interest to many parties.

8. LOCALISED ACOUSTIC EMISSION

Localised AE measurements can be divided into two main regions of activity, namely at the:

- liquid and gas domes (upper part of cargo tank, above liquid level);
- cargo tank walls (lower parts of cargo tank, exposed to liquid).

Global AE is usually first carried out on the liquid and gas domes followed subsequently (if necessary) by the cargo tank walls. When global AE measurements have been undertaken on the liquid and gas domes, localised AE can be carried out on the liquid and gas domes in parallel with global AE measurements being carried out on the rest of the cargo tank.

Localised AE mapping has been carried out by various companies in regions where indications have been detected by global AE. In general, the localised AE mapping has more precisely located each indication and reduced the amount of excavation necessary to expose the defect. As a result, the extent and number of repairs have been able to be minimised and the ship returned to service as quickly as possible. The measurements may also assist in quantifying each indication in terms of its contribution to the VDR as a whole, although GTT's view is that it is impossible to estimate such a contribution.

Localised AE should be undertaken during "quiet" times (e.g. at night, when "noisy" operations on board are minimal; but during daytime may be possible), and when the associated tank IS is under partial vacuum, typically with a differential pressure between the IS and IBS of ~500 mbar. Localised AE can therefore be undertaken during the period when the differential pressure is stabilised subsequent to completing the initial global AE measurements on the tank walls. An example variation of insulation space pressure against time is shown in **Figure 8** below.

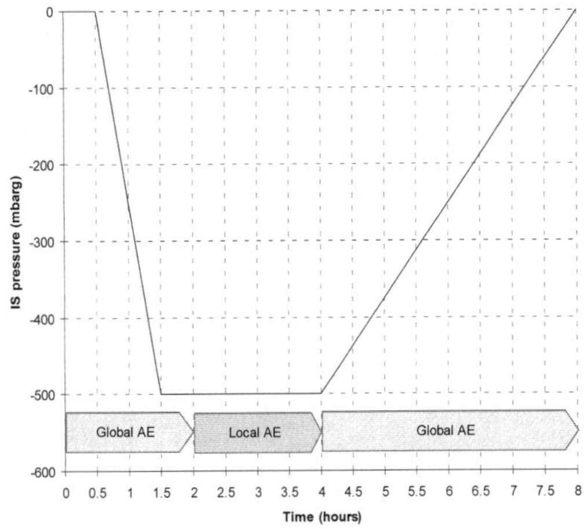

Figure 8: Insulation Space Pressure vs. time - example cycle when both global and local AE are being undertaken

Portable equipment (**Figure 9**) is used to carry out localised AE measurements on the outside, and inside as far as possible, of the liquid dome area, and elsewhere if desired.

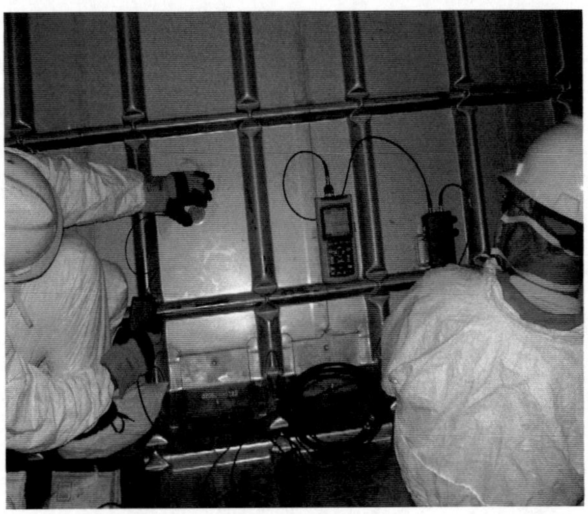

Figure 9: Example portable equipment – recording local AE from inside of cargo tank

Microsoft Excel™ can be used to create local AE mapping of areas with AE values greater than a predetermined threshold. From the local AE mapping, the cargo tank construction drawings, and previous experience, it is possible to estimate the zone of possible damage. From the zone of possible damage, and based on experience from other membrane LNG ships, the likely contribution of any possible localised damage to the overall tank vacuum decay rate (VDR) may be able to be estimated, however GTT's view is that it is impossible to estimate such a contribution.

Hearing tests and global AE may detect and locate indications in way of the tank walls. Using localised AE readings can give more precise location of these indications therefore reducing the amount of excavation required to expose the defect.

9. COMPARISON OF AE APPROACHES

A simplistic comparison of the two approaches is:
- Global AE for CCS – "broadband" in nature, focussing in the range 10 to 100 kHz (centred around 30-50 kHz;
- Local AE for CCS – "narrowband" in nature, focussing on 20 to 40 kHz (centred around ~30 kHz).

A consequence of the differing frequency ranges chosen is the necessary sensor spacing to ensure reliable detection [**Ref. 7**]. A simplistic comparison is:
- Global AE for CCS – sensor spacing typically 2 to 3 m;
- Local AE for CCS –sensor spacing variable typically 0.3 to 0.5 m but dependent on magnitude of signal and resolution desired.

10. THERMOGRAPHIC ASSESSMENT OF MEMBRANE INTEGRITY

TAMI© is a thermal (infra-red) inspection method that has developed in recent years as a test and investigation method in order to allow the direct localisation of potential defects in the secondary barrier during operations (i.e. in service). The test is based on using the thermal gradient of the insulation on a loaded vessel to detect, by thermal imaging, the impact on the inner hull of the gas passing through the secondary barrier. In this test, the pressure of the primary space has to be higher that the secondary space, as shown in **Figure 10** below.

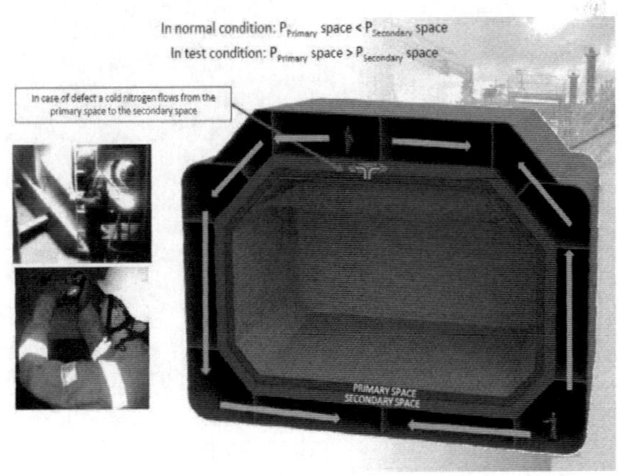

Figure 10: Relative pressures (higher in primary than in secondary space)

The cold gas (blue arrows on **Figure 11**) of the primary space goes through any secondary barrier abnormal porosities and create a thermal indication on the inner hull (**Figure 12**). A thermal camera (**Figure 13**) is used to detect the potential thermal indication on the inner hull.

Figure 11: Flow of gas

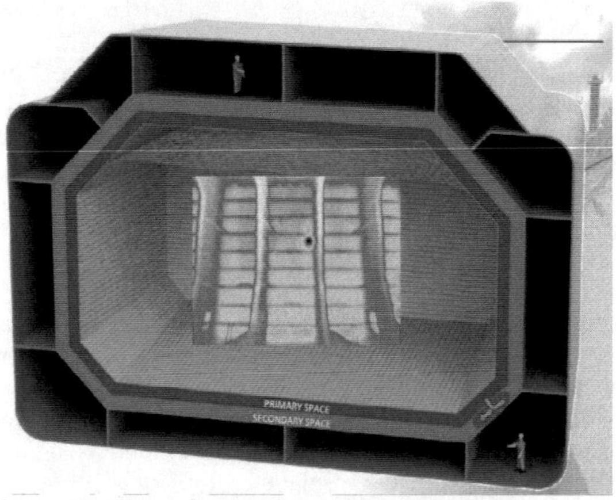

Figure 12: View of tank with one thermal image (showing one spot of interest)

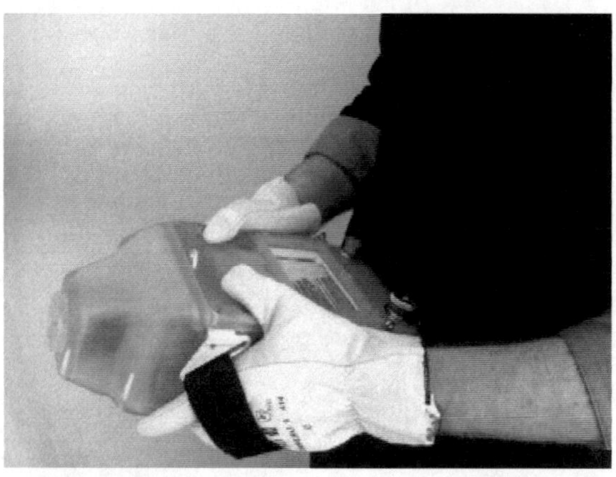

Figure 13: Thermal imaging camera (intrinsically safe)

The images obtained with the thermal camera have to be "post treated". The aim of the post treatment is to determine the origin of the thermal signature observed on the hull. In fact, several heat transfer phenomena take place in the insulation (conduction, natural convection and forced convection). The TAMI© process and the knowledge of the test engineers permit determination of the origin of the cold spot, if any are found.

TAMI tests are generally performed at sea, up to 6 months prior to scheduled dockings, to allow sufficient time to plan the docking in terms of any necessary materials and personnel required. A TAMI test, covering all 4 tanks of a 138-155k membrane LNG ship, might take of the order of 1 week to carry out, assuming that a 4-person test team was deployed. Typically, around 2 days per tank are required, see **Table 1** below.

Table 1: TAMI test – Phases and timings

1st Day		2nd Day	
Phase 1	Phase 2A	Phase 2B	Phase 3
Aim: -To make a reference state for the thermal inspection -To ensure that no cold spot exists on hull in normal conditions	Aim: -To control the temperature of the hull with an inverted pressure regulation	Aim: -Measurement of a temperature gradient	Aim: -To control the state of the inner hull

As regards interpretation of test results, relatively small changes in inner hull temperature are being looked for. Over the four phases of a TAMI test (see **Table 1** above) a significant temperature difference might only be of the order of an increase of 1 degree C, hence a sensitive thermal imaging camera and careful results interpretation are required.

Operators conducting TAMI test should be appropriately qualified, for example certified in accordance with the requirements of "EN 4179: European Standard establishment regarding the qualification and certification of personnel performing NDT".

11. CONCLUSIONS

This paper has considered the assessment of the in-service condition of the secondary barriers of Mk III membrane LNG ship cargo containment systems. For such elements, a combination of LDPT, SBTT, hearing tests, global AE, localised AE and Thermographic Assessment of Membrane Integrity may be used in effective combination to minimise the period that the ship may be out of service during the special survey required by classification and regulatory bodies.

12. ACKNOWLEDGEMENTS

Some of the content of this paper, and a number of the images, have been sourced from material prepared by GTT and Cryovision, two companies well known in the LNG ship world. To both these companies the author wishes to express his thanks for permission to reproduce some of their material. In particular he would like to thank Mr David Colson of GTT and Mr Julian Glory of Cryovision. Note however that any opinions expressed in this paper are those of the author, and should be taken as independent of and not supported by GTT.

13. REFERENCES

1. J M QUENEZ (GTT – GazTransport & Technigaz)) and K H JOH (SHI – Samsung Heavy Industries), "LDPT for Mk III LNGC", *GasTech conference,* 2009.

2. GTT External Document, "Secondary Barrier Tightness Test – Specification M1307*", ED 1136, Rev. 4*, Sept. 2000.

3. GTT External Document, "TAMI (Thermal Assessment of Membrane Integrity) Test", *OPSS-MTD_000034, Rev. 04*, 17/10/2014.

4. GTT External Document, "Acoustic Test Procedure", *ED 3122, Rev. 0*, Feb. 2007.

5. GTT External Document, "Low Differential Test Procedure (LDPT)", *ED 3670, Rev A*, Sept. 2008.

6. CRYOVISION, "TAMI (Thermal Assessment of membrane Integrity) Test Procedure", *OPSS-MTD-00034, Rev. 04*, 17/10/2014.

7. CEN/TC 138, "Non-destructive testing – Acoustic emission", *standard prEN 15857*, 2008.

14. AUTHOR BIOGRAPHY

John R Maguire, FRINA, FIMarEST, FICE, FIStructE, is an independent consultant. He has been employed in a professional capacity in a number of industries since 1976. Prior to working as an independent consultant he was employed by W S Atkins (Engineering and Management Consultants) from 1976 to 1989; by Lloyd's Register EMEA (Consultancy Services) from 1989 to 2012; by MOL (Mitsui OSK Lines – ship owner, operator and manager) from 2013 to 2014; and by the Health and Safety Executive (HSE) from 2015 to 2016. His main areas of work for all 4 companies have been in design, analysis, measurement, investigation and assessment; on land, offshore and at sea.

AN ECONOMIC STUDY OF LNG CARRIERS WITH DIFFERENT CONTAINMENT SYSTEMS ON SPECIFIC TRADE ROUTES

T Lamb, University of Michigan, USA and **R Ramoo**, ALTAIR Engineering, Inc., USA

SUMMARY

An Economic Study of LNG Carriers with different containment systems (including a new one) and a number of typical routes is presented. In addition, the impact of physical constraints, such as Length, Beam and Draft are considered as well as the use of alternate fuels.

NOMENCLATURE

AIP	Approval In Principal
B	Beam
Cb	Block Coefficient
CDTS	Cubic Doughnut Tank System
D	Depth to Main Deck
d	Draft
FLNG	Floating LNG Processing and Storage
GTT	Gaztransport & Technigaz
IHI	Ishikawajima-Harima Heavy Industries
L	Length Between Perpendiculars
LBP	Length Between Perpendiculars
LNG	Liquefied Natural Gas
LOA	Length Overall
MDO	Marine Diesel Oil
MSDSS	Michigan Ship Design Synthesis System
RFR	Required Freight Rate
RPUF	reinforced polyurethane foam
SPB	Self-supporting Prismatic Type B
SHP	Shaft Horsepower
t	Tonnes

1. INTRODUCTION

There are a number of LNG containment tank systems in use, though the two prominent are the Membrane and Spherical tanks. The IHI SPB has been used only on two LNG Carriers years ago, but both ships are still operating. A new (originally invented in 1972 but improved from 2004 to present) containment system is the Cubic Doughnut Tank System (CDTS). Approval in Principal (AIP) has been obtained from American Bureau of Shipping (ABS) in March 2015.

The main purpose of the paper is to examine the impact of different containment systems (including the CDTS) on a number of typical trade routes.

Most current LNG Carriers have been designed with a number of restrictions, mainly Draft but also Overall Length, to suit access to most existing LNG Terminals and Beam to suit Panama Canal. 'No restriction' designs were also developed so that the effect of the restrictions on design, cost and Required Freight Rate (RFR) could be developed. The impact of speed on ship selection for a particular trade route is also examined as well as the use of boil off gas or marine fuel oil for the propulsion of the ships. To accomplish this, a Ship Design Synthesis System was used.

2. ROUTES AND RESTRICTIONS

Eight Cases, covering five routes, based on proposed projects, were examined as well as the effect of dimensional restrictions, speed and alternate propulsion fuel, namely:

Case 1. Maryland USA to India via Suez Canal – 2.8 mtpa
Case 2. Maryland USA to India via Suez Canal – 2.8 mtpa with no dimensional restrictions
Case 3. Maryland USA to India via Cape of Good Hope – 2.8 mtpa
Case 4. Maryland USA to India via Cape of Good Hope – 2.8 mtpa: Impact of Speed
Case 5. Australia to China – 5.3 mpta
Case 6. Australia to Japan – 5.3 mtpa
Case 7. Alaska to Japan- 3.6 mtpa
Case 8. Use of alternate Propulsion Fuel

3. CONTAINMENT SYSTEMS

Three existing containment systems of Spherical, Membrane and IHI SPB along with the CDTS were included in the study.

The Spherical (Moss), Membrane (GTT) and SPB have been around for a long time and are well known. The CDTS is an independent tank containment system that combines the spacial packing efficiency of the membrane system, with the structural efficiency of the curved shells used in spherical tanks, giving a containment system as robust as the IHI SPB system but without the associated manufacturing complexity and cost. The CDTS has been described in detail, including its development history in earlier papers (1-3).

4. DESCRIPTION OF THE CUBIC DOUGHNUT TANK SYSTEM

The marine transport of LNG is a mature technology that is almost 60 years old. In the early years of LNG Carrier design, a number of containment systems were developed and the most successful were the Moss Rosenberg Spherical Tank and the Gas Transport Membrane Lining System. In 1972 a new tank containment system was invented that made high utilization of the hold space as well as the above deck space, unlike the spherical tank, resulting in a significantly shorter ship length for the same cargo capacity. Further development was suspended when the US broke off diplomatic relations with Algeria in 1974 ceasing almost all US based LNG transport operations at that time.

Since 2004 the design efforts have focused on the development of the use of CDTS for both the marine transport (LNG Carriers) and Floating LNG Processing and storage platforms (FLNG).

The use of the CDTS results in the following advantages:

- Significant reduction in ship or platform length for equivalent cargo capacity; 8 to 10% compared to Spherical, 4 to 6% Membrane, OR almost 40% more capacity than Spherical and 25% Membrane and 18% more than SPB for same length,
- Significant reduction in Gross Tonnage (14% and 23% compared to Membrane and Spherical respectively),
- Significant reduction (33%) in the overall construction schedule compared to the membrane tank system,
- No restriction on partial filling of tanks due to sloshing loads,
- Reduced (estimated) cost of LNG carrier by 7% compared to Spherical and 10% Membrane,
- Provided ease of construction and ease of installation in the ship,
- Offered superior structural efficiency,
- Utilized simple support system,
- Better protection from side collision damage – 4m versus 2.5 m for Membrane Ships and 3.25m for Spherical Tank Ships,
- Better protection from bottom damage – 4m versus 2.75 m for Membrane Ships,
- Excellent Boil Off Gas performance – estimated at ~0.13% with 250mm RPUF and ~0.08% with 400 mm RPUF,
- Internal Connectivity provides Natural Conductive Cool-Down Pathways – low pre-loading spray down time with comparable spray-in mass flow-rate,

all combining to offer a cost effective solution for LNG carriers of any size but especially the Ultra Large carriers being considered for the future marine transportation of LNG.

Figure 1 shows isometric views of the CDTS and its installation in a ship's Hull. A comparison of tank outlines is shown in Figure 2 for the spherical, membrane, IHI SPB and CDTS tanks of equal volume. The compact nature of the CDTS is clearly seen. Table I shows the capacity benefit of the CDTS over the other containment systems for the same LOA of 300 m.

5. MSDSS DESCRIPTION

The Ship Design and Operating Economics program was developed in 1965 and used continuously since. In 2003 it was upgraded by a graduate student, M. Kotinis, at University of Michigan (MSDSS) (4). It has been used to assist in the design of tankers, bulk carriers, short sea ROROs, dry cargo ships, fishing trawlers and LNG Carriers (5). Full descriptions of the design and operating approaches have been presented in these earlier papers. The economic design program requires as input Annual Cargo Weight, Stowage Factor, Voyage Round Trip Distance, along with whether there is backhaul cargo available. Along with these is a large amount of both design, operating and economic data that is inserted into the program as tables. A range of speed, hull form coefficient and stability/beam ratios can be selected or single values of each can be used.

6. MSDSS VALIDATION

The MSDSS was validated for LNG Carrier Design by comparing its results with existing ships for the first two containment systems and a concept designs of LNG Carrier for the IHI SPB and CDTS containment systems. The results are shown in Table II and show the close relationship. Additional validation of the MSDSS and its output was achieved by comparison of the results of another recent economic design system paper authored by Dutch and Belgian researchers (6).

From Table II, it can be seen that the predicted optimal principal particulars match well with the actual vessel data. Another area that was validated was the building cost for the LNG Carriers. The published price (what the builder received – may or may not cover builder's costs) had significant scatter (Figure 3) as actual prices vary depending on market conditions. In-lieu of how the shipyards price ships, absolute price estimation is less reliable than the predicted cost differentials between the different containment tank types. While best efforts were made to validate the differences in cost between the spherical and membrane systems, it proved to be difficult without factoring the geographical location of the yards. For ships built in Japan, there was a clear benefit for spherical tank ships over membrane ships. However, in Korean yards, the spherical tank ships were priced slightly higher than membrane ships. This could be attributed to the larger volume of spherical tank ships that Japanese yards build or because of national interests. A similar trend was observed with membrane ships built in Korean yards. In this study, to circumvent, the aforementioned challenges, building cost is estimated using objective measures such as principal particulars, materials and labor input rather than reported **PRICE.**

7. RESULTS

What should be included in an economic design analysis is always a problem. Obviously for a vertically organized company it will own the wells, the loading and unloading facilities and the ship(s). In this case the total system should be optimized. However, for this study only the ship part was examined. The speed for all cases, except Case 4, is 19.5 knots.

7.1 CASE 1 – MARYLAND USA TO INDIA VIA SUEZ CANAL

The designs were restricted to solutions with Beam less than 49 m and a maximum draft of 12.2 m. The designs for each containment tank system are shown in Table III a, b, c, and d. Figure 4 shows the results as Required Freight Rate (RFR) for different number of ships.

Table IV shows the best solution for each of the containment systems. They all require five ships except the Spherical LNG Carrier which requires six. The CDTS has significant reduction in RFR compared to all the other systems with restrictions.

As stated above only the ship part of the complete LNG system was considered and the results will be influenced by the LNG loading and unloading facility liquefaction and storage capacities. When these are included it tends to reduce LNG Carrier capacity, speed and number of voyages per ship per year.

7.2 CASE 2 – MARYLAND USA TO INDIA VIA SUEZ CANAL WITH NO DIMENSIONAL RESTRICTIONS

It can be seen from Figure 5 and comparing Tables III (d) and (e) that removal of dimension restrictions offers two additional solutions. However, the three ship solution is not practical at a capacity of 361,802 CM. The four ship option is feasible at a capacity of 255,127 CM and provides a 11% reduction in RFR compared to the CDTS LNG Carrier solution in Table III (d).

7.3 CASE 3 – MARYLAND TO INDIA VIA CAPE OF GOOD HOPE

The Suez Canal charges fees which are a significant part of the annual operating cost for ships transiting the canal. For a 173,000 CM LNG Carrier it can be $400,000 one way loaded and $300,000 return in Ballast. The results for Cape of Good Hope route are given in Table V (Suez results are repeated for easy comparison) for Membrane and CDTS LNG Carriers. The results surprisingly show a small difference between the RFR for the different routes with the benefit being with the Cape route. Note that the CDTS LNG Carrier can service the Cape route with 4 larger ships whereas the Membrane requires 5 (317,400 CM versus 244,000 CM), and the RFR benefit increases from $5 to $14 for 4 ships instead of 5.

7.4 CASE 4 - MARYLAND USA TO INDIA VIA CAPE OF GOOD HOPE: IMPACT OF SPEED FOR CDTS LNG CARRIER

It is generally well known that for any economic ship study the largest and slower ship will give the best economic performance.

In the case of an LNG Carrier which uses Boil Off gas as fuel, the normal relationship is not as pronounced, as clearly shown in Figure 5. The crosses show designs within the 300 M LOA restriction. It is interesting how the RFR for all these solutions are within a range of 127 to 141 $/ton with the six ships at 22 knots, surprisingly, giving the lowest RFR.

Figure 6 shows the results when diesel fuel oil is used for propulsion and the normal effect of RFR increasing with increasing speed can be observed. From Figure 6, with all fuel oil propulsion, and when there is no LBP restriction, there is no bottoming out of the RFR curves within the speed range considered, though for seven ships at 14 knots it is close. There are no solutions for fewer ships with lower speeds as the required LNG Capacity would be too large.

Figure 7 shows the LBP for a range of speeds and number of ships for the use of Boil Off gas as the propulsion fuel on the delivery voyage. When the 285 M LBP restriction is applied, the solutions are six ships for 22 knots, seven ships for 20 and 18 knots, and eight ships for 16 and 14 knots

7.5 CASES 5, 6 AND 7 – OTHER TRADE ROUTES AND AVAILABLE ANNUAL CAPACITIES

The round voyage distance for Australia to China and Japan is about the same at 7000 and 8000 nautical miles respectively, so the results are almost the same. The speed was constant at 19.5 knots. The distance from US to Japan is 9000 nautical miles. The best solutions for LNG Carriers using the CDTS are shown in Table VI.

7.6 CASE 8 – USE OF ALTERNATE PROPULSION FUEL

The topic of this Case has already been discussed in Case 4 in order to show how the fuel use impacted selection of best LNG Carrier size and speed for a given trade route. Today most LNG Carriers are being ordered with duel fuel diesel engines for propulsion. Existing LNG Carriers use the LNG boil off as fuel for marine boilers and steam turbine propulsion; recent deliveries and LNG Carriers on order mostly have the duel fuel engines. The final case to be examined is for the same Route as Case 1 but with the additional cost of complete marine fuel oil for propulsion. In the case of the full marine fuel oil there is a choice of installing a reliquifaction plant on the LNG Carrier or to simply flare off the boil off gas at sea. This choice is the alternative shown in Figure 8 but it may not be available much longer as environmental laws may prohibit flaring off gas.

The annual lost opportunity for additional income for each voyage and each ship, and then fleet annual income, from selling reliquefied gas, can be derived by multiplying the difference in RFR by Boil Off in tonnes and the Number of Voyages per year which for each of the 5 ship solution would be around $300,000. Then the investment for the reliquefaction plant and the annual operating cost of the reliquefcation plant would have to be lower than that to be economically viable.

8. CONCLUSIONS

The study clearly shows that each project (mtpa and round trip distance) has a unique combination of number of ships, size and speed that will give the lowest RFR. Clearly there is no LNG Carrier Size or Speed that fits all potential trade routes.

Acquiring an LNG Carrier and finding it to be the best size and speed for a specific trade route, would be most unlikely. Using a 'standard' (say 173,000 CM, 19.5 knots) LNG Carrier for a specific route probably will not be an exact match and will require the ship to change speed to match required (or available) annual LNG Delivery Capacity. While it is quite common to de-rate a diesel engine to suit lower speed it is at an unnecessary cost. It is better to design each LNG Carrier for the exact annual cargo and select the number of ships, LNG capacity, speed and engine to match required SHP, which gives the best economic performance.

For a given trade route, the smallest number of ships, largest capacity, and slow speed will give the best economic performance and thus competitive position. However, when other supply chain and/or ship restrictions are considered, it will impact the best solution, probably increasing best speed and decreasing best size.

The paper also shows the usefulness of the MSDSS for such economic design studies.

Finally, it is suggested that the predicted performance of the CDTS is of sufficient benefit to warrant further consideration by operators and shipbuilders of LNG Carriers.

9. ACKNOWLEDGEMENTS

The authors would like to acknowledge with thanks the support of ALTAIR Engineering and their vision of a future for the CDTS. The application of their advanced CAE tools to a practical idea has enabled the CDTS to be further developed and proven its feasibility.

10. REFERENCES

1. LAMB, T, and KOTINIS, M, 'A Set-based Ship Design Synthesis System,' *(2003) IMDC 2003*, Athens, Greece

2. LAMB, T, 'The Effects of Intended Trade Route on the Optimum Size of LNG Tankers,' *(1974) LNG-4 Conference, Algeria*, June 27, 1970

3. LAMB, T, and RAMOO, R, "The Application of a New Tank Containment System to ULTRA-Large LNG Carriers', *OTC* 2009

4. LAMB, T, and RAMOO, R., 'A New Concept for CNG Carriers and Floating CNG/Oil Processing and Storage Offshore Platforms,' *CNG Forum*, London 2009

5. LAMB, T, and RAMOO, R, 'A New Tank Containment System for Floating LNG and CNG Processing and Storage Offshore Platforms,' *OTC* 2010

6. GRIN, R., and van de VOORDE, E, 'Weather-Related Economics of Natural Gas Transport for Two Propulsion Plant Configurations,' *RINA Conference, Design and Operation of Gas Carriers*, 2004

11. AUTHORS BIOGRAPHY

Mr. Lamb has over 60 years' experience in the design and construction of many ships of different types and spent his last 11 active years teaching and performing research at University of Michigan retiring in 2006. He served a five-year apprenticeship at H.M. Dockyard Rosyth and graduated from Durham University in 1958 with his B.Sc. in Naval Architecture. He is a Fellow of RINA, and SNAME, and was the Editor of and 5-chapter contributor to the 2003 edition of the SNAME SHIP DESIGN AND CONSTRUCTION book as well as being the SNAME Ship Design Committee Chairman for many years. He is a Professional Engineer in the U.S., Chartered Engineer in UK and Eur Ing.

Mr. Ramoo, Vice President of Engineering at Altair ProductDesign has over 30 years' experience in the use of computational methods for design ideation and Advanced Structural Analysis. Mr. Ramoo graduated with a Diploma in Marine Engineering from the Singapore Polytechnic in 1977 and holds a MSc. in Mechanical Engineering from The Ohio State University. He has broad experience in using advanced CAE methods in structural design of land vehicles and marine structures.

TABLE I – DESIGN COMPARISONS FOR RESTRICTED LENGTH, BEAM AND DRAFT FOR THE FOUR CONTAINMENT SYSTEMS

For LBP 285 M Beam 49 M and Draft 12 M RESTRICTION				
	Membrane	Spherical	SPB	CDTS
LBP m	285	285	285	285
Beam m	49	49	49	49
Depth m	28	28	34	28
Draft m	12	12	12	12
Displacement t	126,541	121,070	146,018	151,935
Lightship t	36,779	41,000	50,800	38,500
Lightship/Displacement	0.291	0.339	0.348	0.253
CARGO DEADWEIGHT t	89761	80070	95218	112823
CARGO CAPACITY c m	186653	166500	198000	234607
Relative Capacity to Membrane	100	89	106	126
Relative Capacity to Spherical	112	100	119	141
Relative Capacity to SPB	94	84	100	118
DHP MW	33500	33000	36000	40000
DHP/CAPACITY	0.179	0.198	0.182	0.170
Overall Volumetric Efficiency *	0.477	0.426	0.417	0.600

The Overall Volumetric Efficiency is derived by dividing The LNG Capacity by the surrounding dimensional envelop, that is:

Overall Volumetric Efficiency = LNG Capacity/ (L x B x D x Cb)

TABLE II – MSDSS VALIDATION

	MSDSS VALIDATION							
	MEMBRANE		SPHERICAL		IHI SPB		CDTS	
SHIP CHARACTERISTIC	Actual	MSDSS	Actual	MSDSS	Actual	MSDSS	Design	MSDSS
LBP M	262	262	277	277	226	226	250	250
Beam M	43.4	44	48.4	48	40	40.1	44	44.8
Depth M	26	26	26.5	27	26.8	27	26	25.91
Draft M	12.12	12	11.5	11	11	10.67	11.5	11.28
Displacement t		101176		109529		71200	100350	100339
Lightship t		31026		34166		21940	30414	29930
Deadweight t	70150	71050	75000	75363	48917	48935	69936	70409
Cargo Deadweight t	65750	65710	70975	71100	45417	45501	65410	65911
Cargo Capacity CM	138122	138141	142700	142703	88995	89200	135000	135900
Speed Knots	19.5	19.5	19	19	18.5	18.5	19.5	19.5
SHP kW	30260	29878	27000	26983	21000	20740	28358	28746

TABLE III – 2.8 MTPA MARYLAND TO INDIA VIA SUEZ CANAL

(a)- FOR MEMBRANE TANK LNG CARRIERS

MEMBRANE		NUMBER OF SHIPS				
		5	6	7	8	9
LBP	M	283.5	268.0	256.1	246.0	237.8
BEAM	M	46.7	44.4	42.5	41.0	39.6
DEPTH	M	28	27	26	25	26
DRAFT	M	12.0	11.3	10.8	10.4	0.9
DISPLACEMENT	t	131,256	109,740	94,677	83,484	83,484
DEADWEIGHT	t	93,773	76,731	64,916	56,273	49,695
CARGO CAPACITY	CM	191121	156521	132959	115569	102199
SHP	kW	36,049	32,510	29,793	27,640	25,870
BUILDING COST PER SHIP	$	221 M	198 M	182 M	168 M	158 M
RFR	$/t	115	124.5	133.4	141.8	150

(b) - FOR SPHERICAL TANK LNG CARRIERS

SPHERICAL		NUMBER OF SHIPS				
		5	6	7	8	9
LBP	M	293.3	277.7	265.9	255.8	247.6
BEAM	M	49.0	46.3	44.2	42.5	41.0
DEPTH	M	28	27	26	25	24
DRAFT	M	11.9	11.3	10.8	10.0	9.7
DISPLACEMENT	t	141,334	118,642	102,772	87,585	78,573
DEADWEIGHT	t	93,828	76,702	64,931	56,319	49,688
CARGO CAPACITY	CM	190717	156635	133007	115931	102506
SHP	kW	37,727	34,096	31,290	29,663	27,742
BUILDING COST PER SHIP	$M	226	203	185	163	152
RFR	$/t	118.4	128.6	138.1	147.3	155.4

(c) - FOR IHI SPB TANK LNG CARRIERS

IHI SPB		NUMBER OF SHIPS				
		5	6	7	8	9
LBP	M	279.9	263.1	254.0	241.2	231.7
BEAM	M	45.9	43.8	41.8	40.5	129.2
DEPTH	M	32	31	30	29	28
DRAFT	M	12.3	11.5	10.9	10.5	10.2
DISPLACEMENT	t	127,776	106,495	91,889	80,719	72,151
DEADWEIGHT	t	93,746	76,718	64,938	56,288	49,664
CARGO CAPACITY	CM	188459	154349	131169	113969	100813
SHP	kW	35,467	32,005	29,207	27,163	25,373
BUILDING COST PER SHIP	$M	233	208	190	175	163
RFR	$/t	119.0	128.3	137.8	145.8	153.6

(d) - FOR CDTS TANK LNG CARRIERS

CDTS		NUMBER OF SHIPS				
		5	6	7	8	9
LBP	M	271.3	257.0	246.3	236.3	228.0
BEAM	M	46.1	43.7	41.7	40.2	38.8
DEPTH	M	28	27	26	25	24
DRAFT	M	12.1	11.4	10.9	10.5	10.1
DISPLACEMENT	t	129,973	108,654	93,871	82,766	74,174
DEADWEIGHT	t	93,772	76,677	64,948	56,306	49,681
CARGO CAPACITY	CM	191263	156821	133327	115937	102579
SHP	kW	45,155	40,356	36,743	30,184	28,165
BUILDING COST PER SHIP	$M	198	177	162	150	140
RFR	$/t	110.7	120	128.7	136.7	144.4

(e) - FOR CDTS TANK LNG CARRIERS WITH NO RESTRICTIONS

CDTS	NUMBER OF SHIPS					
	3	4	5	6	7	8
LBP	344.8	308.8	284.8	266.8	252.4	241.2
BEAM	58.5	52.4	48.4	45.3	43.1	41.2
DEPTH	30	29	28	27	2	25
DRAFT	13.2	12.3	11.6	11.1	10.8	10.5
DISPLACEMENT	228,133	164,286	131,915	109,634	94,322	82,996
DEADWEIGHT	170,611	121,141	94,002	76,714	64,926	56,320
CARGO CAPACITY (	362171	255602	197662	160658	135690778	117374
SHP	49,486	41,651	36,498	32,752	29,981	27,757
BUILDING COST PER SHIP $	302	245	210	186	168	155
RFR	93.7	103.2	111	120	127.8	135.1

TABLE IV – COMPARISON OF DIFFERENT CONTAINMENT TANK SYSTEMS FOR US TO INDIA VIA SUEZ CANAL

2.8 mtpa Maryland to India via Suez		CONTAINMENT TANK SYSTEM			
Length Overall Restricted to 300 M		Spherical	Membrane	IHI SPB	CDTS
Number of Ships		6	5	5	5
LBP	M	277.7	283.5	279.9	271.3
BEAM	M	46.3	46.7	45.9	46.1
DEPTH	M	27	27	32	28
DRAFT	M	11.3	12.0	12.3	12.1
DISPLACEMENT	t	118,642	131,256	127,776	129,973
DEADWEIGHT	t	76,702	93,773	93,746	93,772
CARGO CAPACITY	CM	156635	191121	188459	191263
SHP	kW	34,096	36,049	35,467	45,155
BUILDING COST PER SHIP	$	203	221	233	198
RFR	$/t	128.6	115	119.0	110.7

TABLE V – COMPARISON OF SUEZ AND CAPE OF GOOD HOPE ROUTES RFR

NUMBER OF SHIPS	4	5	6	7	8	9
Membrane Suez		115	124.5	133.4	141.8	149.8
Membrane Cape of Good Hope		105.9	114	121.8	129.2	136.2
CDTS Suez		110.7	120	128.7	136.7	144.4
CDTSCape of Good Hope	96.9	105.3	113.1	120.5	127.5	134

TABLE VI – SOLUTIONS FOR AUSTRLIA TO CHINA AND JAPAN AND US WEST COAST TO JAPAN

CDTS		AUS TO CHINA 5.3 mtpa		AUS TO JAPAN 5.3 mtpa		US TO JAPAN 3.6 mtpa	
Number of Ships		3	4	3	4	2	3
LBP	M	305.2	276.2	321.6	286.6	327.7	275.0
BEAM	M	51.9	47.0	55.7	48.8	55.9	46.9
DEPTH	M	30	29	31	30	32	29
DRAFT	M	11.5	10.2	11.8	10.8	13.1	11.6
DISPLACEMENT	t	152744	107956	178189	124201	203641	122544
DEADWEIGHT	t	116736	79143	136182	91910	160229	92576
CARGO CAPACITY	CM	238572	162732	288955	193517	341821	195536
SHP	kW	39688	32435	43504	35219	46710	35019
BUILDING COST PER SHI	$M	212	173	238	189	249	180
RFR	$/t	31.9	35.9	35.5	38.8	36.5	41.5

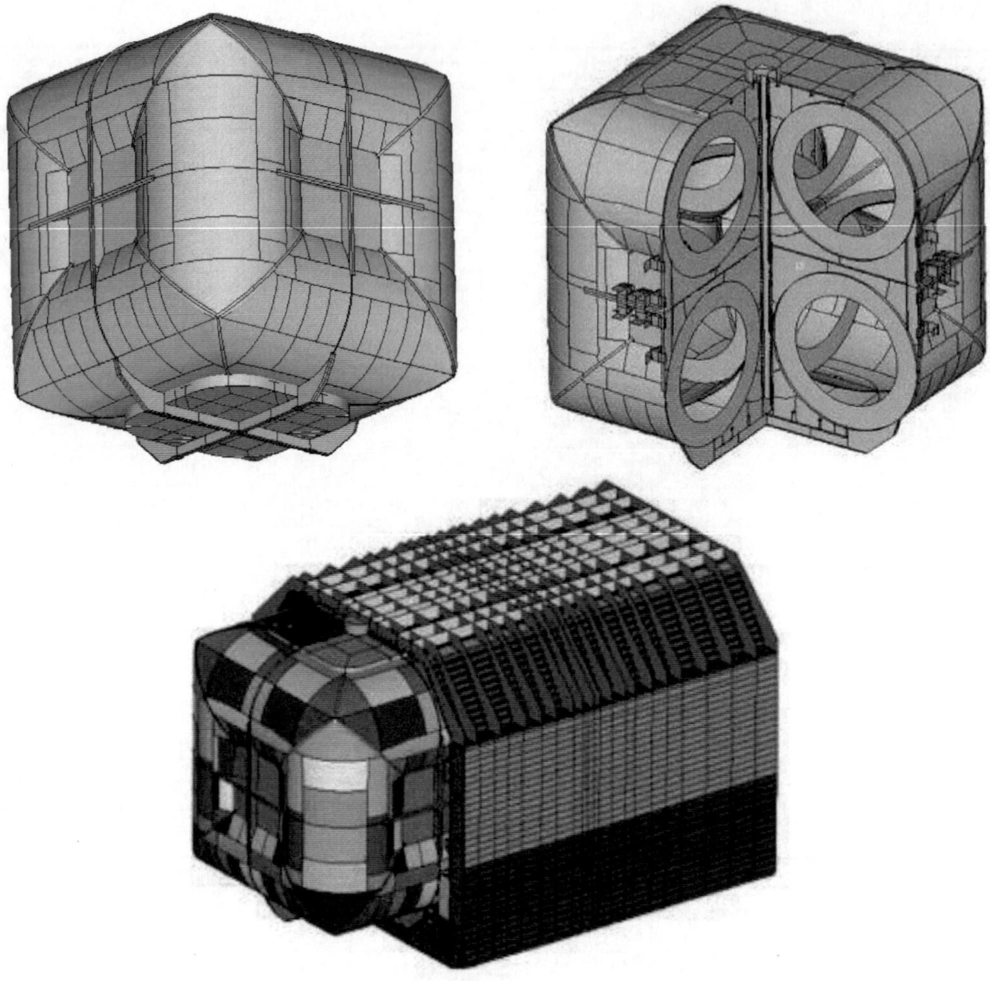

Figure 1: Isometric Views of CDTS

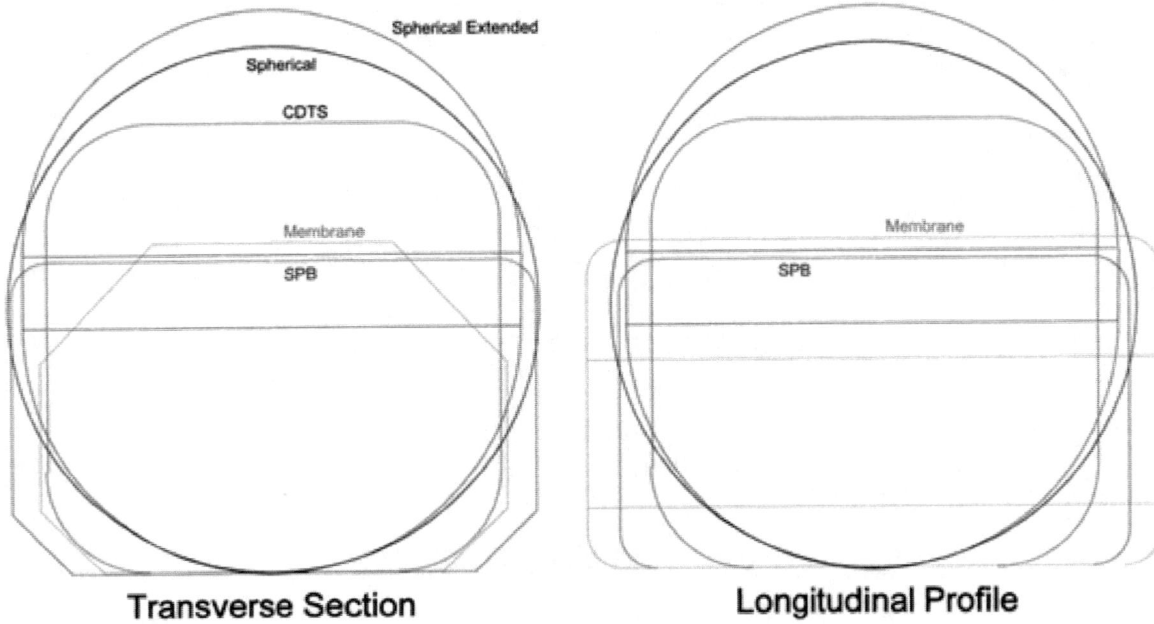

Figure 2: Comparison of Outlines for the Four Containment Systems for Equal Volume

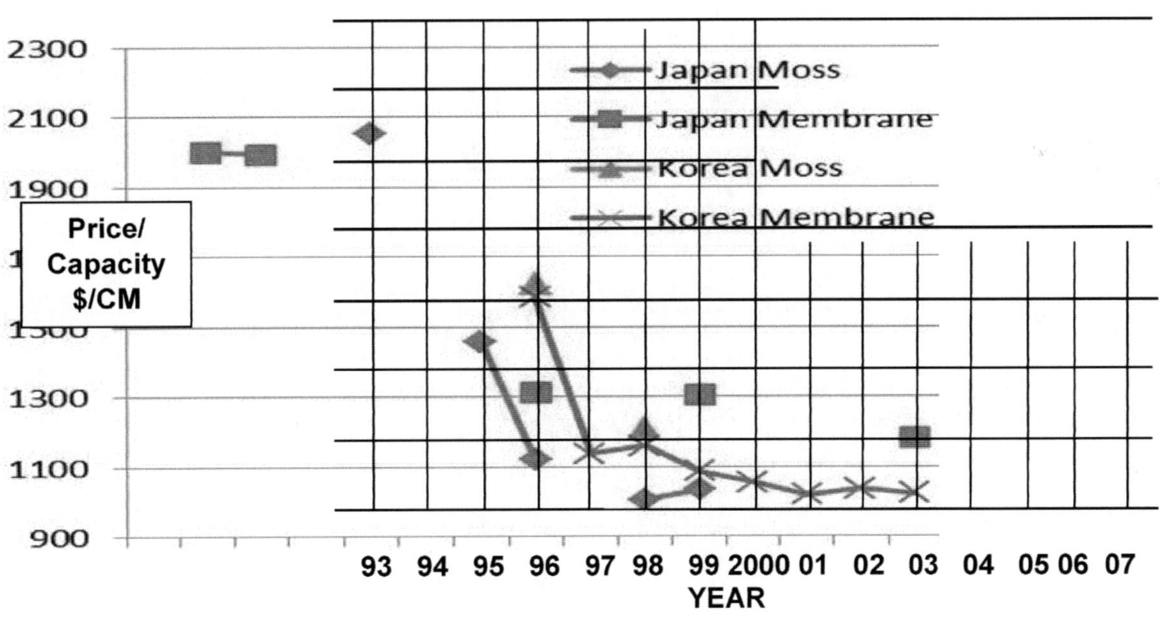

Figure 3: LNG Carrier Price/Capacity

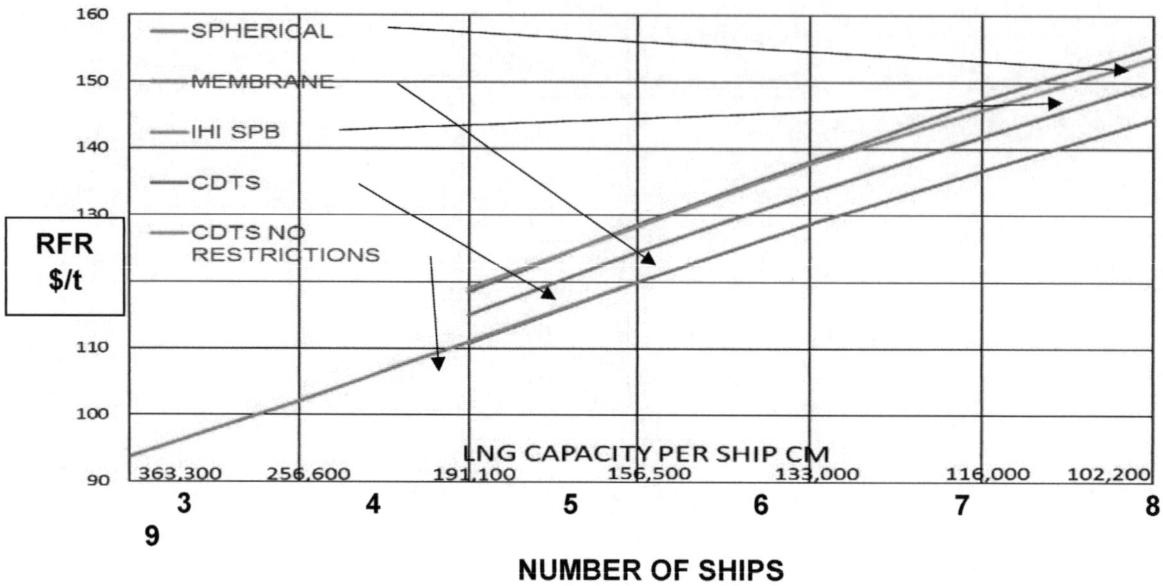

Figure 4: RFR for Maryland to India via Suez Canal for Different Containment Systems and Number of Ships

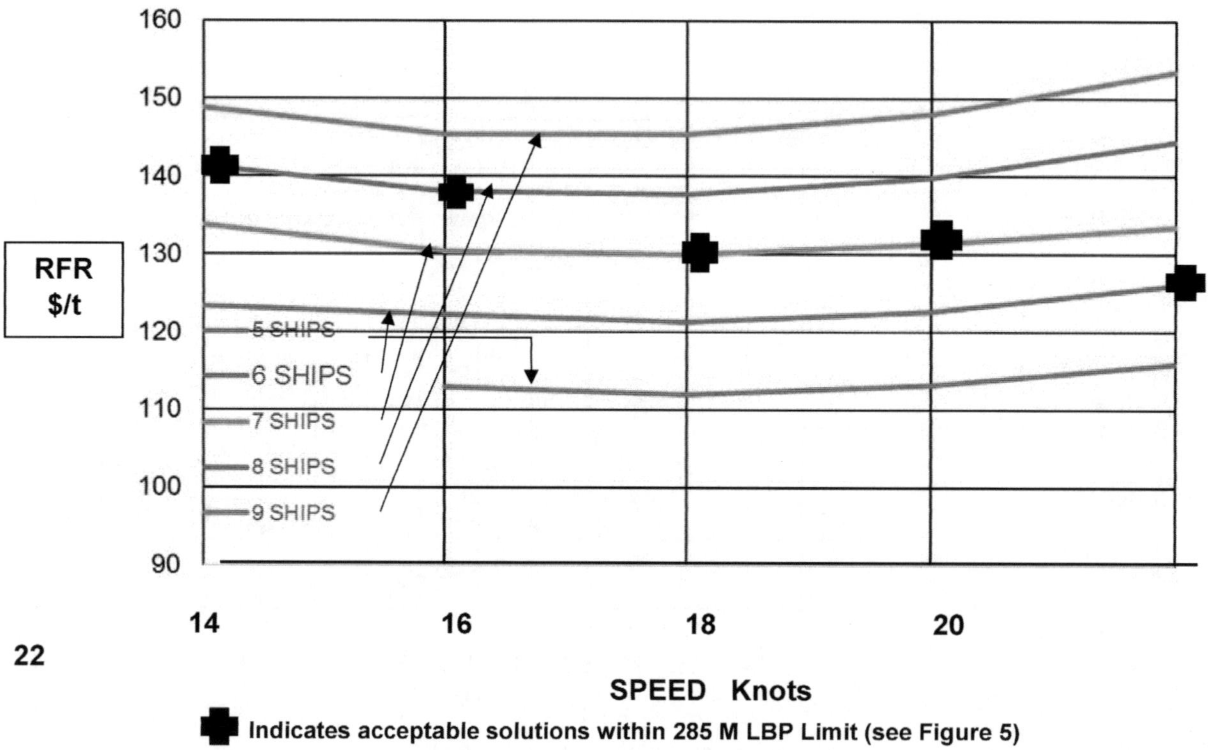

Figure 5: Impact of Speed on Economic Performance: CDTS Cape of Good Hope Using Boil Off for Propulsion Fuel

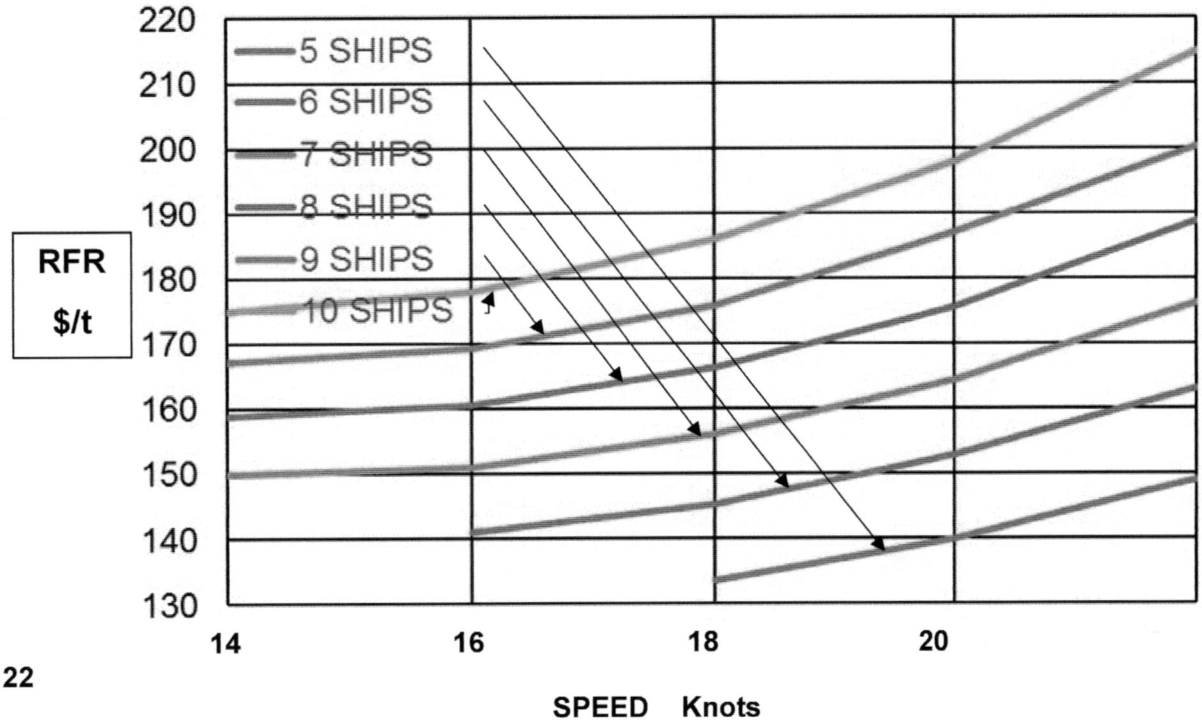

Figure 6: Impact of Speed on Economic Performance: CDTS Cape of Good Hope Using Diesel Fuel for Propulsion

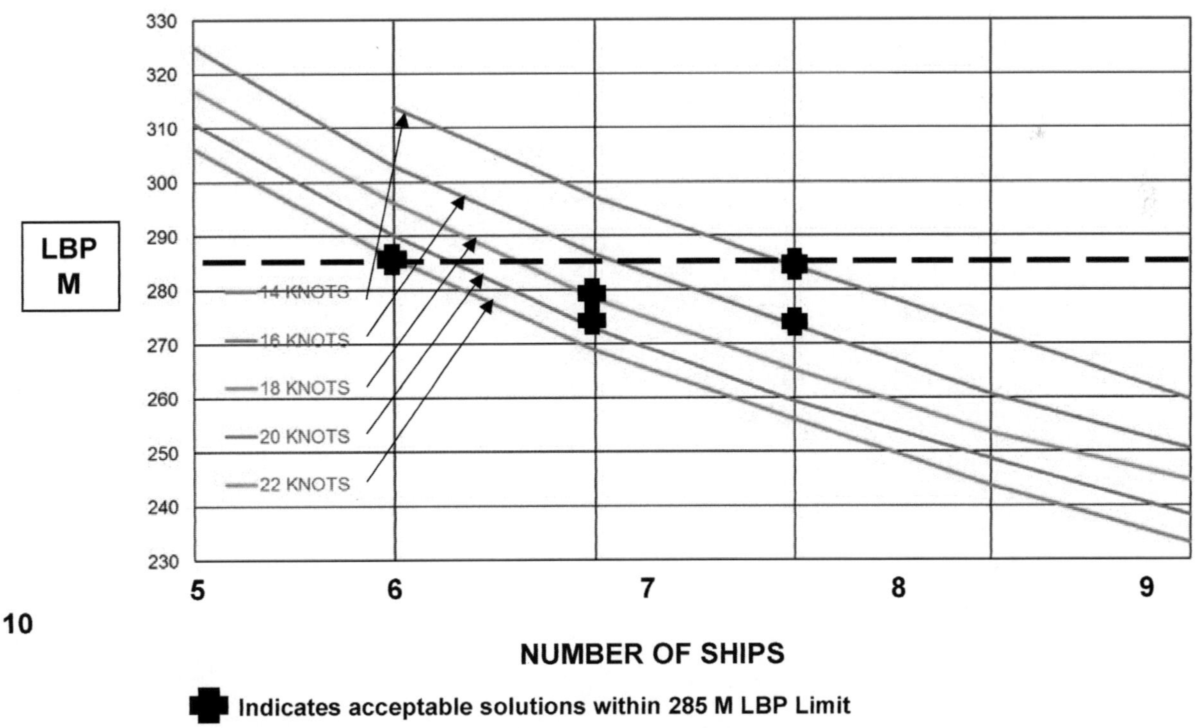

Figure 7: Length Overall/Speed Relationship: CDTS Cape of Good Hope

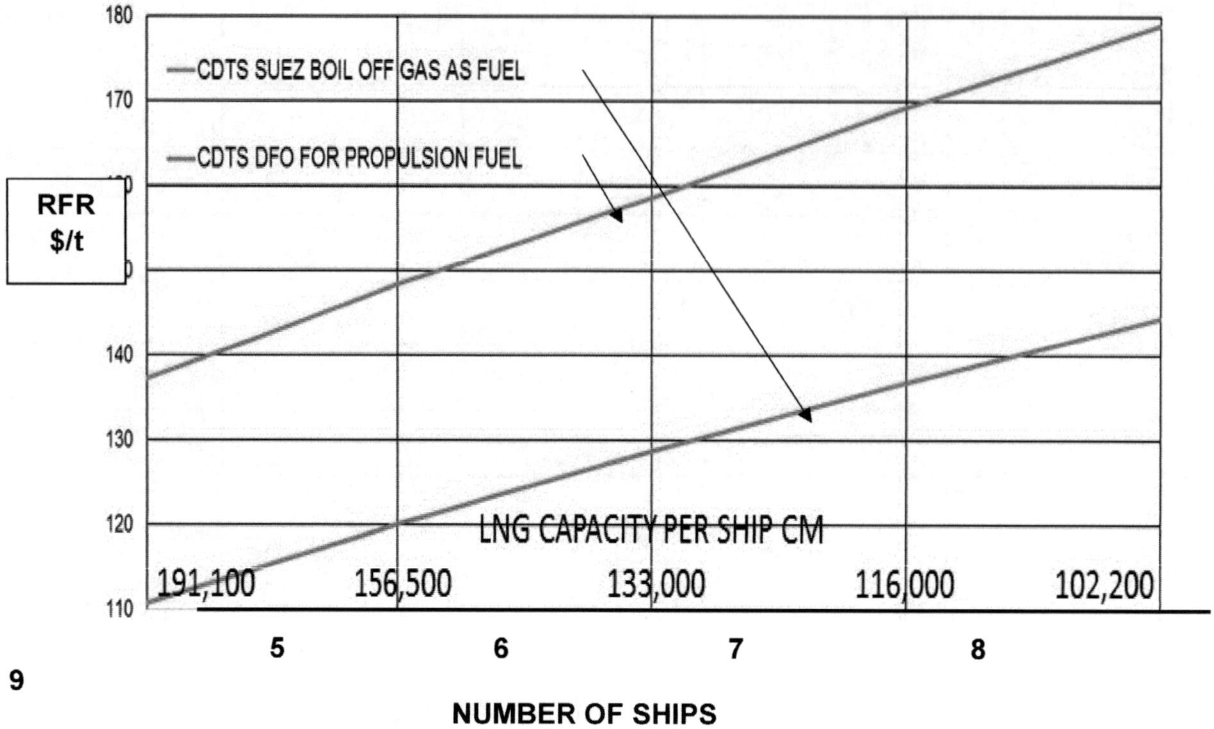

Figure 8: Comparison of Fuels for CDTS Suez Case

RE-THINKING LNG TRANSFER SYSTEMS USING CRYOGENIC FLEXIBLE HOSES

V Lagarrigue, Trelleborg Oil & Marine

SUMMARY

This paper outlines the development and qualification of a cryogenic flexible hose that enables to re-think LNG transfer systems. Primarily designed for LNG transfer in tandem configuration on Floating LNG projects, the cryogenic floating hose opens new frontiers in small scale LNG market segments. It becomes in particular a key component for cost effective alternative to standard LNG receiving terminals.

1. INTRODUCTION

Moving LNG production to an offshore setting presents a demanding set of challenges. Along with the design and construction challenges of developing FLNG facilities, extra thought has been required regarding the transfer of the product; LNG transfer must now battle with the effects of winds, waves, and currents in open seas. In environments where conditions are much more demanding, conventional marine loading arms will not suffice as they could result in a shut-down of the liquefaction plant in bad weather conditions.

2. CRYOLINE LNG HOSE – A DEDICATED AND PROVEN DESIGN FOR LNG TRANSFER APPLICATIONS

Trelleborg has been designing a floating cryogenic hose which will ensure LNG offshore transfer with minimum Boil-off-Gas (BOG) generation, using a combination of high flexibility, reliability, and long service life. In addition, the new Cryoline LNG hose seeks to meet LNG operator and contractor's offloading requirements, with particular emphasis on safety, flowrate capacity, and operation availability. Therefore, the following parameters were put forward for the development of the hose:

- Inner diameter: 6 to 20 inch (150 to 500 mm)
- Maximum Allowable Working Pressure > 15 barg
- Design temperature range: -321°F / -196°C to +104°F / +40°C
- Minimum Bending Radius (MBR): < 3 m
- Same deployment as conventional oil marine hoses

However, while offshore LNG facilities have an anticipated 25 year operating lifetime, the nominal operating life of a cryogenic hose is five years, during which it is expected to be submitted to a maximum of 500 offloading operations (assuming an offloading operation of twice a week).

The end result of the development process was the Trelleborg Cryoline LNG floating hose, which is based on a hose-in-hose concept that consists of a field-proven outer rubber marine hose with an inner LNG composite hose, already well established, in particular for use in LNG ship-to-ship transfer. As Trelleborg has been manufacturing and supplying composite and rubber-bonded hoses for many years, the design utilizes existing and proven technologies, and is a combination of industry-leading expertise in both composite and rubber-bonded hoses. More specifically, this new floating cryogenic hose is made up of several key components, including an inner cryogenic hose, an outer protective hose, an efficient insulation layer and an integrated leak monitoring system, as shown on Figure 1 below.

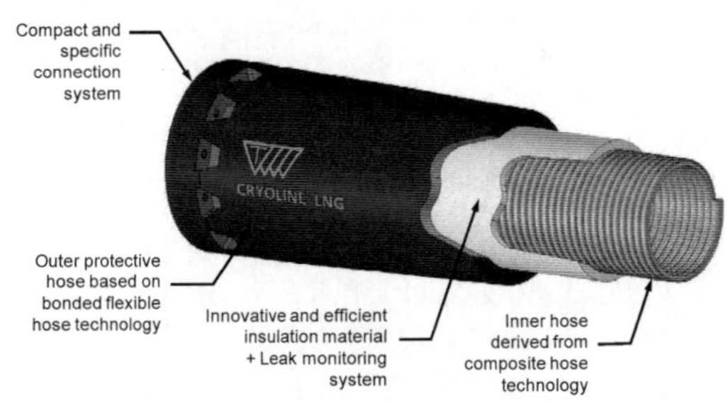

Figure 1: Cryoline LNG hose key components

The inner cryogenic hose has been derived from the latest developments in composite hose technology, best known for its high flexibility and proven suitability for LNG ship-to-ship transfer in side-by-side configuration. Composite LNG hoses usually consist of multiple unbonded, polymeric film and woven fabric layers, trapped between two stainless steel wire helices, which gives the hose its convoluted shape, one being internal and one external.

Broadly, the film layers provide a fluid-tight barrier to the conveyed product, while the woven fabric layers provide the mechanical strength of the hose. The number and arrangement of multiple polymeric film and woven fabric layers is specific to the hose size and application. The polymeric film and fabric materials are selected to be compatible with the conveyed product and the extreme operating temperatures. Composite LNG hoses have already proven their suitability for such an application as this technology has been validated through many full scale static and dynamic tests, and many offshore ship-to-ship LNG transfers.

The outer protective hose has been developed based on flexible rubber bonded hose technology, best known for its high resistance to fatigue and high ability to comply with harsh environmental conditions. Rubber bonded hoses are usually made of rubber, steel rings, steel cable reinforcement layers and end-fittings. Typically, the hoses come in sections of 39 feet / 12 meters, which are bolted together in situ, either in a horizontal or vertical assembly configuration. They are often either used for floating or submarine configurations, and conform to OCIMF or API 17K specifications. Such a technology offers major technical and project execution advantages, making it a very attractive solution, especially from the perspective of overall offloading terminal optimization.

To demonstrate the fatigue life of rubber-bonded hoses, Finite Element Analysis (FEA) tools have been specifically developed during the qualification program to compute the stress within any steel wires or the strains within rubber layers. This technology offers high resistance to fatigue as the rubber bonded hoses might include a tapered Integrated Bending Stiffener (IBS) at each end, progressively decreasing the stress concentration in the flange's connection.

Also, as the steel parts of the flange are fully integrated within the rubber, the system is corrosion free, and due to the high insulation performance of rubber (thermal conductivity about 0.3 W/m.K), rubber bonded hoses inherently show high thermal insulation properties.

The annular space between the inner and outer hose is filled with innovative and efficient insulation materials which have excellent properties over the full range of temperatures (from ambient to cryogenic temperatures). As long as external environmental conditions are above +41°F / +5° C, the insulation layer is designed so that no ice will form on the outer cover of the cryogenic hose.

These materials have been designed to reduce heat loss within the structure, to protect the outer rubber bonded hose from cryogenic temperatures and to ensure LNG hose buoyancy. Derived from advanced technologies, they consist of an open structure, which enables the operator to inert the annular space with nitrogen, between each offloading operation. Moreover, they also have excellent resistance properties to dynamic loads (fatigue) generated by the application.

An integrated leak monitoring system based on optical fiber technology for gas leak detection has been included in the design in the annular space between the inner and outer hoses. The two optical fibers – for redundancy purpose – are wound around the composite LNG hose to check the evolution of temperature within the structure and prevent any abnormal activity during the offloading operation. The operator will benefit from a fast, effective and reliable control system to monitor the offloading conditions.

Finally, a compact and specific connection system has specifically been designed for the application. This new technology will typically consist of 39 feet / 12 meter long sections, which will be connected together – either onshore or offshore – with threaded rods and nuts, in the same way that conventional flexible bonded hoses for oil application would be.

As such, leading manufacturers have also developed a new concept of end fitting to ensure load transfer and leak tightness, and to minimize heat loss within the offloading lines. The design of the connection system includes dedicated seals for cryogenic application which are used for static and dynamic applications, exhibiting excellent sealing integrity in gas and fluid applications, and withstanding rapid changes in temperature.

The concept of the end fittings has been developed through Finite Element Analysis (FEA) to take into account coupled thermal and mechanical loads at the very first steps of the design process. In a second step, the calculations have been validated through full scale tests performed on a dedicated test bench to validate the design of the connection system, to demonstrate the tightness of the connection design at ambient and cryogenic temperatures, and to endorse the choice of the cryogenic sealing technology. For example, cyclic compression loads up to 200 tons have been applied on a full scale connection, highlighting a safety factor of 10 in service conditions on the key components.

3. QUALIFICATION OF AN LNG FLOATING HOSE

In 2008, a specific European Standard EN 1474 [2] was developed by the industry to give a common frame for the qualification of new LNG transfer systems and flexible hoses. The main challenge for this LNG tandem offloading system qualification is to qualify the LNG transfer floating hose according to the EN1474-2 standard, which gives general guidelines for the design, material selection, qualification, certification, and testing details for LNG transfer hoses for offshore transfer in floating configurations. Such hoses need to be durable when operating in the marine environment and flexible with a minimum bending radius compatible with the handling and operating requirements of the transfer system.

Figure 2 below shows the overall philosophy for the development and qualification of the Cryoline LNG hose. The ongoing qualification of the floating cryogenic hose is a significant challenge in itself and will enable Trelleborg to receive a Type Approval Certificate (TAC) – issued by a classification society – in accordance with the European Standard EN 1474-2.

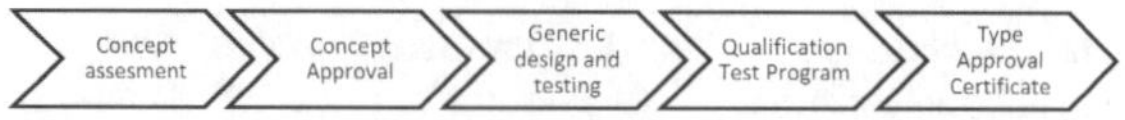

Figure 2: Cryoline LNG development and qualification philosophy

A qualification test program (QTP) has been therefore proposed by Trelleborg according to the European

Standard EN 1474-2 requirements, which comprise a complete set of full-scale tests. Typically, laboratory testing on materials have to be performed to ascertain that mechanical properties are suitable for the intended application, and that components' performance are not adversely affected by exposure to environment – both internal and external, e.g. LNG, NG or marine environment – over the hose design life. Typically, the following tests are required within the qualification process:

Mechanical properties of all materials at ambient and cryogenic temperatures

Effect of LNG exposure on all materials which are exposed to cryogenic temperature

Effect of ambient climatic exposure on all materials which are exposed to atmosphere

Effect of marine environment, e.g. corrosion of metallic components.

In addition, the qualification test program includes 24 destructive or non-destructive tests – either static or dynamic – in ambient or cryogenic conditions to demonstrate the cryogenic hose's ability to operate in the foreseen configuration, e.g. capability of withstanding temperature/pressure cycles, capacity of operating at the maximum flow rate, and fatigue resistance to concurrent tension and bending cycles.

Tests	B	C	G	F	E	H
Leak test	X	X	X	X	X	X
Leak test	X			X		
Bending test		X	X*		X*	
Bending test		X				
Tensile test		X				
Tensile test				X		
Torsion test		X				
Torsion test				X		
Pressure cycle test		X				
Thermal and pressure cyclic test					X	
Burst test						X
Burst test	X					
Water flow test		X				
Bending cyclic fatigue testing			X*		X*	
Electrical testing	X	X	X	X	X	X
Crush testing & Impact testing				X		
Characterization of buoyancy				X		

Figure 3: Cryoline LNG hose qualification program test matrix

Figure 3 presents the qualification test matrix, demonstrating the distribution of the qualification tests – both in ambient and cryogenic conditions – that will be performed on the full scale hose prototypes. To do so, six prototypes have been manufactured and tested according to EN 1474-2 standard recommendations. Except the cryogenic burst test and the full scale water flow test – which have been done at TNO facilities in the Netherlands – all the qualification tests have been performed at Trelleborg facilities in France, under the survey of classification society, Bureau Veritas.

Figure 4 below shows examples of qualification tests: both bending tests at ambient and cryogenic temperatures, as well as tensile and torsion tests at room temperature.

Figure 4: Qualification tests on 20" Cryoline LNG hose

A cryogenic burst test was conducted on a 20" Cryoline LNG hose in May 2015 at TNO facilities in the Netherlands. The burst pressure occurred at 106 barg (objective was 100 barg) and it is important to note that the external rubber envelope was not damaged during the test (see figure 5 below).

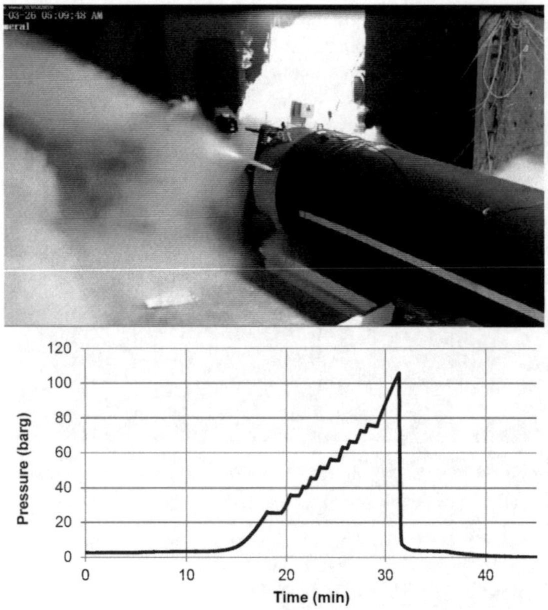

Figure 5: Cryogenic burst test (TNO, May 2015)

A dedicated fatigue test bench has been installed in Trelleborg facilities to conduct dynamic cycling tests on full scale prototypes. The test bench enables the application of static and dynamic tension loads, static and dynamic bending loads, as well as static torsion loads on 20" Cryoline LNG Hose prototypes (see figure 6 below). The hoses are filled with liquid nitrogen (LN2) during the test, and regularly pressurized at Maximum Allowable Working Pressure (MAWP). A specific data acquisition system provides the ability to monitor and record test parameters, through various sensors signals, e.g. temperature, pressure, loads, position etc. The test configuration includes two hose sections connected together to check the connection's ability to withstand recurrent dynamic loads of the same magnitude than in normal and extreme operating conditions.

The fatigue test will be completed on full scale prototypes – including a complete connection system – to prove that the Cryoline LNG withstands recurrent dynamic loads for long service life. Typically, the qualification test program includes bending cyclic conditions that reproduces the dynamic load conditions that the cryogenic hose will be subjected to in service conditions.

Figure 6: Cryogenic fatigue test bench (bending configuration)

The qualification of the 20" Cryoline LNG hose according to EN1474-2 standard was successfully completed in April 2016 with the issuance of a Certificate of Compliance from Bureau Veritas.

4. TANDEM LNG OFFLOADING SYSTEM

The tandem offloading system detailed in Figure 6 is designed to store, deploy, connect and disconnect the floating hoses, while guaranteeing an optimal level of control, safety and maintainability. The three floating hoses have an internal diameter of 20" and a length of approximately 200m for a nominal distance between the LNG terminal and the LNG carriers of 100m (typical value that may be adjusted to the need and philosophy of each operator). Two hoses are dedicated to LNG transfer, targeting total offloading flowrates of 10,000 to 12,000m3/h. The third hose is dedicated to vapor return.

The three hoses end terminations are bolted to a common connection head that includes the termination valves and a by-pass line, with a valve between the two liquid lines together. A stainless steel tubular structure protects the piping components and houses floatability modules that make the connection head float when transferred from the LNG terminal to the LNG carrier.

The hoses' storage system is located at the stern of the LNG terminal. The three hoses are stored independently from one another on horizontal stages rotating around the same shaft structure. Coming out of the rotating stages, the hoses are guided down along the aft transom of the LNG terminal (see figure 7 below).

A storage and maintenance platform for the connection head is installed on the aft transom of the LNG terminal. The connection head and its components can be checked and inspected on this platform before and after each offloading operation. The platform also provides equipment to inert the hoses and to facilitate the connection of the lines used by the LNG carrier to drag the connection head and hoses to the bow loading platform on the carrier.

Figure 7: Tandem offloading system main components

On the LNG carrier side, a bow loading platform is used to lift and connect the connection head and hoses to dedicated manifolds in sea states with significant wave height (Hs) up to 3.5 m (see figures 8 and 9). The connection head can stay connected and be disconnected in Hs up to 4 m. The weight of the connection head is supported by the structure of the platform, so that the couplers connected to the connection head end termination valves see very limited loads. Therefore, standard QC/DC couplers and valves can be used for the purpose. The platform also houses mooring equipment if the operator intends to moor the LNG carrier to the LNG terminal by means of a hawser, even if Total's point of view is to prefer DP2 LNG carriers.

Figure 8: Hoses reel storage system and the connection head maintenance & storage platform

Figure 9: Connection head during lifting operation

The design of the tandem offloading system has been submitted to a qualification of the items that were presenting some novelty (usually existing components used in a new application) and an exhaustive engineering design validation. The qualification has been validated by Bureau Veritas, which awarded the system with a Concept Approval Certificate, in accordance with EN1474-3 and BV NR 542 standards, in 2013.

If the market demands such a system for loading an FSRU, some few additional studies will be performed but nothing insurmountable has been identified thus far. One can also imagine innovative LNG transfer using a floating hose to negate the cost of a traditional jetty.

5. SOLUTION FOR SMALL SCALE LNG TRANSFER

Import and export of LNG is a market in strong growth. The challenge during the last 30-40 years has been that it is very expensive to construct a jetty-based LNG terminal. Such a marine terminal usually consists of a long jetty (often 1-3 km) + mooring arrangement for large LNG ships (300 m long) with associated LNG loading equipment (Loading Arms). In addition, many terminals may also need to construct a large Breakwater Structure (typical 1 km long or more) to reduce the waves, as shown in figure 10 below. Such high capital investment is often unaffordable for small LNG terminals , which is why the LNG industry is currently looking for less expensive and more flexible LNG import and export solutions.

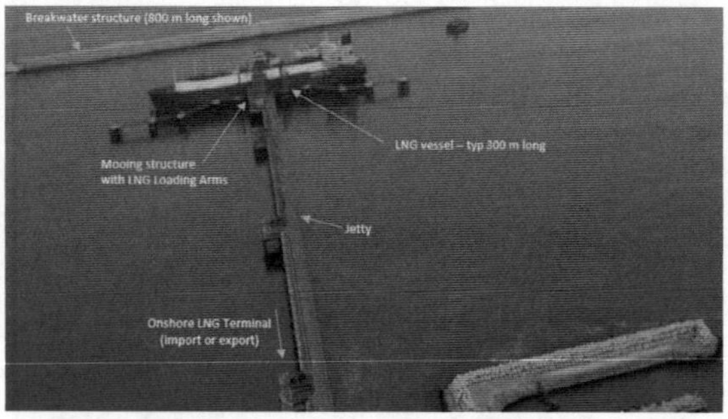

Figure 10: Example of typical LNG Import/Export Terminal

The Autonomous Transfer System – ATS has been developed with the motivation of solving the challenge described above. The cost for the complete ATS will be significantly lower than the cost of a traditional jetty-based, fixed LNG terminal.

Furthermore, the ATS also makes it possible to safely offload LNG carriers at higher wave heights than traditional terminals due to the flexibility of the Trelleborg LNG hoses from the ATS to shore, and the favorable motion characteristics of the semi-submersible barge. The typical operational limit for the ATS system is in the magnitude of Hs 2.0 m, while traditional jetty type terminals typically can accept maximum wave heights beyond Hs 1.0 m.

An overview of a typical ATS system is given in figure 11 below (further details of proposed ATS based on given project parameters is given in the following sections).

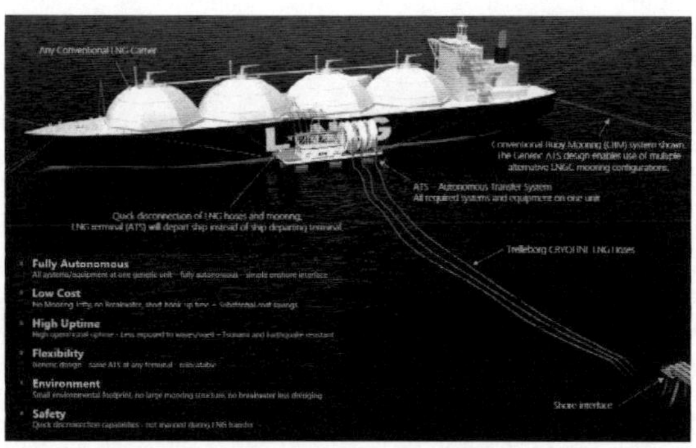

Figure 11: Autonomous Transfer System – ATS

The ATS concept enables the possibility to include all required equipment on the ATS unit. I.e. a complete system can be built at a cost efficient shipyard, be fully tested, and then simply towed to the final location.

The hoses from the ATS have a simple interface to the onshore pipelines by standard flange connections. The complete ATS system can also be relocated to another location if the LNG terminal should stop its operations. The ATS is a fully generic design and can be used at any terminal, worldwide.

As the system for LNG export and import is completely identical, the ATS can be used for both applications. It is also viable for any standard LNG vessel of any size, with no modifications for the LNGC's required.

The system can be installed offshore by use of an submerged insulated LNG Pipe in Pipe from the shore and out to the selected location. The alternative is to construct a LNG pipeline bridge as shown in figure 11 above.

6. PRESSURE DROP ANALYSIS

To demonstrate the flow performance of cryogenic hoses, Total, Saipem and Trelleborg decided to carry out LNG and water flow tests to assess the pressure drop due to characteristics of the flexible hose. This is a significant point of interest for the LNG industry when using floating cryogenic flexible hoses in an offshore LNG transfer application. It is critical to evaluate pressure drops within the LNG offloading lines to properly design the LNG cargo pumps onboard the FLNG, as well as the BOG compressors onboard the LNG carrier. It should be noted that there is no general rule or consensus among the industry on this specific matter. Therefore, the consortium considered it important to develop a global philosophy regarding pressure drop and to launch a feasibility study with the main objectives of accurately understanding the physical principles at stake, and developing a reliable and robust model for pressure loss evaluation.

Steady state 3D simulations on a small scale cryogenic hose with water were initiated at the first phase of the project, using the commercial CFD code STAR-CCM+ by CD-ADAPCO. Extensive numerical simulations have been performed to develop the tool for the modeling of water and LNG flows for the case of small and full scale hoses. The roughness of the inner wall of the hose (fabrics and stainless steel) has also been taken into account, as it was expected to strongly affect the pressure losses. The goal was to obtain the evolution of the pressure drop and the corresponding friction factor with the Reynolds number. To guarantee validity, this part of the project was conducted in close collaboration with the Mechanical Laboratory of Lille (LML) in France.

In particular, cryogenic flexible hoses are tested to assess pressure drops by performing LNG and water flow tests. The approach consists of performing LNG and water flow tests at medium scale and water flow tests at large scale, and then extrapolating the results to assess the pressure drop for LNG flow in cryogenic hose at full scale (ID 20"). The reduced scale hoses are insulated with the same material and identical insulation value (U-value) as the full scale hose.

To achieve this, the following methodology has been defined (as shown on the figure 13 below):

1. Numerical model of a water flow on a 20" hose. The model will be fitted with the results of the water flow test made on a 20" hose in the same conditions. This enables us to better reproduce the internal geometry of the inner hose and, in particular, ensures the ability to carefully define the roughness of the inner hose. Moreover, this step also enables us to validate the conclusions drawn during the previous numerical studies (in particular the choice of the turbulence model to be applied). The fitted model will be also used to evaluate the behavior of the friction factor for flowrates of water above 6,000m3/h and thus for Reynold numbers higher than 5x10-6. The friction factor is also derived from the flow test.

2. Numerical model of a water flow on a reduced scale hose. The model will be fitted with the results of water flow test made on reduced scale hose. This step allows us to accurately and numerically reproduce the internal geometry of the inner hose for the reduced scale hose. This step also allows us to validate the similarity rules between a full scale and a reduced scale hose. This then enables us to extrapolate results from existing results at a given scale to another scale.

3. Numerical model of a LNG flow on a reduced scale hose. This phase will allow us to validate if the pressure drops anticipated for LNG based on tests performed with water are valid and if a simple analytical model can be used. The numerical model for LNG flow defined in the previous phase will be calibrated with LNG flow test results performed on reduced scale hose. It will also be possible to define similarity rules between fluids. At this stage, it will be possible to highlight thermal effects, as well as possible fluid phase evolution. However, thermal aspects are highly anticipated to have a limited impact on pressure drops for some reasons: LNG is subcooled before entering in the hose, the heat losses within the hose are low because of the performing insulation properties leading to almost constant an homogeneous temperature along the hose (about 0.1 °C temperature difference between inlet and outlet).

4. Numerical model of a LNG flow test on a 20" hose. As no full scale test will be performed with LNG, the numerical model is based on the work done during the previous steps. For this phase, the geometry validated with the 20'' water flow test and the similarity rules defined in the previous test are used. Thermal aspect as well as possible phase evolution are also considered in this case.

To extrapolate the results and observations from the small scale flow tests to the full scale and to compare the findings between different fluids (water and LNG), it is necessary to retain the same fluid (water or LNG) velocity between full and small scale, and the same Reynolds number between water and LNG flow tests for the same hose, as shown in figure 12 below.

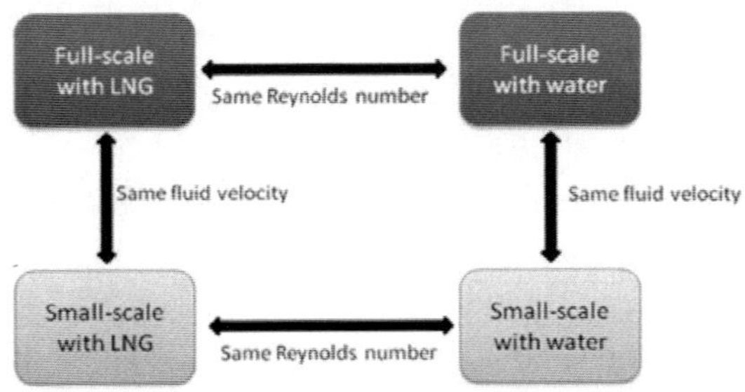

Figure 12: Pressure drop assessment philosophy

The reduced scale hose will have an internal diameter (ID) of 10" maximum. The length of the hose will be in the order of 10 meters (minimum 40 × ID). The length of straight sections downwards from and upwards of the flexible hose location should be at least 10 times the diameter of the hose.

A full scale water flow test was conducted on a 20" hose in June 2015 at TNO facilities in the Netherlands (see figure 13). Flow conditions of up to 6,000 m3/h were tested successfully at inner pressure of up to 8 bara (see figure 14 below).

Figure 13: Full scale water flow test (TNO, June 2015)

The pressure drop in the inner hose was found to be very close to the predicted model, based on classic rules. Figure 14 compares the flow test results against the predictive model response.

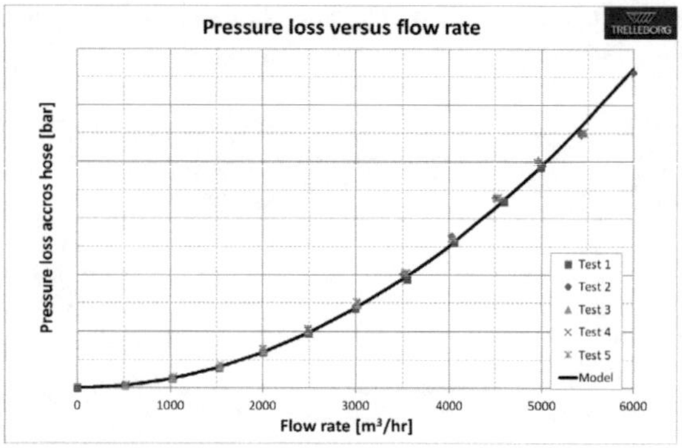

Figure 14: Pressure loss assessment (tests results vs. model response)

This ambitious program started in 2013 and is expected to last a couple of years. The main objective is to build a reliable and robust CFD model, enabling the definition of the pressure drops within an LNG offloading line for a wide range of diameters (at least for diameters ranging from small scale and full scale hoses), as well as for different fluids (i.e. different LNG compositions).

7. CONCLUSION

This paper has explored why and how – in the absence of mature solutions – Trelleborg has been developing and qualifying a 20" floating cryogenic flexible hose for LNG offshore transfer, for both FLNG and small scale configurations.

By enabling nearshore or offshore transfer of LNG, the cryogenic floating hose will pioneer a step change in the safety of this critical operation. This system will also allow FLNG or small scale LNG projects to be considered in harsher conditions, without excessive downtime due to offloading system availability, and with significantly reduced risk. Trelleborg's combined efforts in the development of an innovative tandem offloading system with floating hoses is underpinned by proven technologies and existing industry practices and procedures. Safety, robustness, and simplicity of operations and installation have driven the design of each component.

The qualification of the system is now fully finalized through the full-scale testing of the 20 inch / 50 cm Cryoline LNG floating hose. This thorough development, which combines the expertise of technology providers, suppliers, contractors and a classification society, as well as the support of a major Oil company, will provide the LNG industry with a reliable and robust solution for LNG tandem loading operations.

Derived from existing and proven technologies, the Cryoline LNG floating hose will be a key component of offloading systems for future FLNG or small scale projects. Based on flexible bonded hose technology, which is suitable as an external hose for the floating LNG hose-in-hose concept, the cryogenic hose development program has been focused at an early stage on key elements such as a composite hose suitable for transfer in cryogenic conditions and dedicated end fittings for such applications. Subsequently, those elements have to be integrated within the flexible bonded hose to design a homogeneous, safe and reliable cryogenic hose able to meet European Standard EN 1474-2 prerequisites, as well as operators' offloading requirements.

8. REFERENCES

1 European Standard 1474: Installation and equipment for liquefied natural gas – Design and testing of marine transfer systems
- Part 1: Design and testing of transfer arms
- Part 2: Design and testing of transfer hoses
- Part 3: Offshore transfer systems

2 Guide to Manufacturing and Purchasing Hoses for Offshore Moorings (GMPHOM), including considerations relating to hose system design - Fifth Edition, 2009.

9. AUTHORS BIOGRAPHY

Vincent Lagarrigue is in charge of promoting Oil & Marine complete product range (Oil and Gas transfer systems), including solutions for Floating LNG and Small Scale LNG applications. He has been responsible for the design and qualification of flexible bonded hoses for offshore crude oil transfer (Oil Offloading Lines, Reel systems) and LNG transfer systems.

THE OPTIMUM CONFIGURATION OF A SELF-STANDING TYPE B LNG TANK USING ADVANCED STRUCTURAL ANALYSIS AND SIMULATION TOOLS

R Ramoo, Altair ProductDesign, USA, **B W Eng**, **S H Wu**, Altair Engineering, Ltd, Taiwan, **T Lamb**, University of Michigan, USA

SUMMARY

The paper describes the application of computer aided engineering methods using Altair Engineering's HyperWorks suite of advanced structural design, analysis and optimization tools to develop the design of a Type B tank LNG containment system. A combined utilization factor is used as a yardstick to measure the aggregate of the spatial efficiency of different containment shapes and their maximum structural efficiencies that can be realized. The optimal structural efficiencies of different shapes is assessed using state-of-the-art optimization techniques like Topology and Free-size optimization which is widely used for design ideation and optimization in industries where a high premium is placed on weight savings. In-depth structural analyses per IGC code are then performed on the proposed containment shape to show the feasibility of proposed tank configuration.

NOMENCLATURE

ABS	American Bureau of Shipping
AIP	Approval-In-Principal
B	Beam
CAE	Computer Aided Engineering
CCS	Cargo Containment System
C_b	Block Coefficient
CDTS	Cubic Donut Tank System
CM	Cubic meters
D	Depth to Main Deck
d	Draft
FEA	Finite Element Analysis
IGC	International Gas Code
IHI	Ishikawajima-Harima Heavy Industries
L_{BP}	Length Between Perpendiculars
LNG	Liquefied Natural Gas
L_{OA}	Length Overall
LBF	Leak Before Failure
SPB	Self-Supporting Prismatic Type B
t	Tonnes

1. INTRODUCTION

The two prominent LNG cargo containment systems are the membrane and spherical tank system. The membrane systems typically comprise of a low thermal expansivity liner on the inside of the vessel's hull. A sandwich comprising of a secondary barrier, and insulating framework or layers provide the necessary liner support and thermal insulation to minimize heat ingress into the CCS. This system is not independent nor self-supporting but provides high utilization of the hold space in the ship. The spherical tank system is a cryogenic grade aluminum or nickel steel sphere wrapped with an insulting layer. The containment tank sits on a cylindrical skirt in the hold space. A third LNG containment system is the IHI-SPB which has been used on two LNG carriers that are still in operation. Several new orders for this system have been recently reported. The SPB tanks are independent like the spherical tank CCS but make high utilization of the hold space like the membrane system.

Since the membrane CCSs are not independent and rely on the hull to provide the structural stiffness to mitigate hydrostatic loads, they tend to have the lowest tank weight. However, not being independent limits the volume of above deck space that these tanks can protrude into for a variety of reasons such as hydrostatic loads, the need for a doubled walled deck to prop up the liner, and the need to reduce the free surface area to limit sloshing loads.

While a spherical shape containment system provides the highest structural efficiency when it comes to mitigating hydrostatic loads, these shapes do not pack well in a prismatic hull. As the spherical tanks are independent, they can protrude above the deck and thus provide a somewhat competitive volumetric efficiency to membrane tanks. A more recent variant to the spherical tank is the stretched version wherein an equatorial cylindrical ring is added to limit the diameter of the spherical tanks at higher containment volumes. Another variant also limits the required height by the use of toroidal shaped shells above and below the equatorial cylindrical ring.

The SPB tanks are prismatic and thus have a high below deck space utilization. Although they are independent and self-supporting, they do not protrude above the deck. This is partly attributed to taller tanks requiring thicker plates and more reinforcement at the base of the tank to mitigate the accompanying hydrostatic load. Any gains in volumetric efficiency by increasing the height of the tanks would thus be negated by the higher tank weights required.

A new (originally invented in 1972 but improved from 2004 to present) containment system is the Cubic Donut Tank System. Approval-In-Principal (AIP) has been obtained from American Bureau of Shipping (ABS) in

March 2015. The higher combination of spatial efficiency and structural efficiency of a CDTS tank as summarized in Table 1 warrants further consideration by operators and shipbuilders of LNG carriers. The first part of this paper examines the spatial efficiencies and structural efficiencies of these containment systems in aggregate. The paper than outlines a summary of some of the structural analyses done on the CDTS tank per IGC code using ABS rules.

2. SPATIAL EFFICIENCY OF CONTAINMENT SYSTEMS

The four containment system shapes are superimposed in Figure 1. The spherical and CDTS tanks are superimposed on a layout of a typical membrane carrier in Figure 2. The tanks are of equal volume to the baseline membrane carrier shown in the figure. The spherical tank arrangement would require the vessel to be of a slightly larger beam but comparable length. The spherical tank ship does not require cofferdams but do need some clearance between the tanks and compartment bulkhead resulting in a ship of comparable length. Since the SPB tanks will generally not protrude above the deck for stability considerations and/or to limit the hydrostatic loads at the base of the tanks, the resulting SPB ship length would be also be comparable.

A CDTS tank being constructed primarily out cylinders can manage higher hydrostatic load than a prismatic tank fabricated out of flat plates for the same tank wall thicknesses. This allows for the CDTS tank to be taller than SPB tanks but need not be as high or wide as spherical tanks for the same volume. This higher tank volume on a smaller hold space results in a higher volumetric efficiency as the space above the main deck is utilized more efficiently than the other containment systems.

At a system level this results in a shorter and more compact ship. Since ship length is one of the major drivers of ship cost, the CDTS ship will cost less than other LNG carriers of equal capacity. The cascading benefits of a shorter ship such as lower hull weight is well understood by ship designers. Another benefit of the compactness of CDTS ship is a lower Gross Tonnage. CDTS carriers have the highest capacity to gross tonnage ratio.

3. LIGHTSHIP WEIGHT EFFICIENCY OF LNG CARRIERS

The containment tank's spatial efficiency dictates the length, beam and depth a vessel needs to be to carry a certain volume of LNG but it does not completely factor the vessel's displacement that is required to carry that weight of LNG. The required displacement is a function of the deadweight and lightship weight. Since the preceding section considered equal volumes of LNG, the four types of ships would have comparable deadweights. The lightship weight comprises of the weight of the hull and appurtenances, and the weight of the containment tanks. The weight of the hull is closely correlated to the size of the ship and hence the spatial efficiency; a longer and larger ship would require more plating and stiffeners.

The next factor that needs to be considered is the weight of these different CCSs. Altair's Topology Optimization technology was used to develop the optimal CDTS tank layout which has been described in detail in previous publications in references (1-3). Altair's HyperWorks Free-size optimization was used to evaluate the theoretical lowest weight each of the three independent containment systems can be designed to while meeting code. Practical manufacturing and fabrication requirements generally require less discrete partitioning of the tank shell structure resulting in higher than the theoretically predicted weights. The tank weight of a membrane CCS is difficult to isolate since the hull structure itself provides the structural stiffness to contain the hydrostatic loads attributable to the LNG cargo. The additional structural reinforcement to the hull required to contain these could be apportioned to tank weight but since these loads are coupled to the hydrostatic and hydrodynamic sea keepings loads in deployment, they are excluded for tank weight comparison purposes.

4. OPTIMIZATION RESULTS

Topology optimization was used to develop the layout of the optimal placement and orientation of the stiffeners in the CDTS tank. The objective function in Altair's topology optimization is the minimization of structural compliances when subjected to the combination of loads. A sample topology optimized result for the bottom cover is shown in Figure 3. This preferred material distribution was used as an input to ideate the closeout cover design as shown in Figure 4. A similar approach was used to develop the rest of internal arrangement of the CDTS CCS.

A finite element model of the topology optimized 40K CM CDTS tank was then used to compare the weight of the CDTS CCS against other optimized containment systems and benchmarked against widely reported values. Free-size optimization is best used after the topology optimized finite element model has been developed to obtain the optimal plate partitioning strategy. Finite element models of a spherical and SPB tanks were also developed based on published dimensions scaled to a 40K CM sized tank. Altair's Free-size Structural optimization with imposed stress constraints were used as a basis to compute the "best case" shell thicknesses in the tanks required to match the structural compliance of the spherical tank. The membrane tank weight was estimated from published data.

The structural compliance (C) used for Topology and Free-size optimization is based on the following relationship:

$$C = \frac{1}{2}U^T F \qquad with \qquad KU = F$$

or

$$C = \frac{1}{2}U^T KU = \frac{1}{2}\int \varepsilon^T \sigma dv$$

For a structure with applied force (F), the compliance (C) can be considered a reciprocal measure of the stiffness (K)

$$C = \frac{1}{2}U^T F = \frac{1}{2}\frac{F^T F}{K^T} = \frac{1}{2}\frac{f^2}{K},$$

$$with \qquad \frac{1}{2}f^2 = constant$$

For maximum stiffness (K), the compliance (C) can be minimized. For a structure with an applied displacement (U), the compliance (C) can be considered a direct measure of the stiffness (K)

$$C = \frac{1}{2}U^T F = \frac{1}{2}U^T KU = \frac{1}{2}\frac{u^2}{K},$$

$$with \qquad \frac{1}{2}u^2 = constant$$

The thickness and stress distributions are shown in Figures 4 and Figure 5. Once the required thicknesses were evaluated, the comparative weights indices for the different tanks were calculated and reported in Table 2.

With the comparative tank weights established, these were then scaled up for a 170K carrier to establish the lightship weights of LNG ships using the different containments systems. The Michigan Ship Design Syntheses (MSDSS) program was used to predict the optimized ship dimensions for the different type of LNG carriers. Some published data of similarly sized vessels is included as a reference in Table 3.

5. COMBINED UTILIZATION EFFICIENCY FACTOR

The volumetric efficiency of the containment system quantifies how well the design is laid out to package the largest LNG cargo volume on the smallest footprint and hold volume, quite analogous to how an automotive or aerospace engineer would like the largest usable space in an automotive body or fuselage architecture with the least amount of material. The lightship weight efficiency measures how well the available structural material is used to contain the cargo.

Spherical containers invariably have the highest structural efficiency since the hydrostatic loads are mitigated in a purely membrane state. This is however not reflected at a system level in the lightship weight efficiency due to the larger ship size required to house the tanks. Cylindrical shapes also predominantly mitigate in a membrane mode via hoop stresses except for the close-out and end caps. Prismatic structures are the least efficient for mitigating hydrostatic as the loads are resisted in a bending mode thus requiring a significant amount of structural reinforcement for the same containment volume. This is seen in the higher tank weight required for the same structural compliance values from the OptiStruct Free-size results in Table 2.

The aggregate of the volumetric efficiency and the lightship weight efficiency provides a measure of the utilization efficiency of the various containment systems. Another benefit of a high spatial efficiency is the compactness of the carrier which is reflected in the lower Gross Tonnage measure.

6. STRUCTURAL ANALYSIS ON CDTS CCS

The next sections of this paper summarizes some of the structural analyses done on the CDTS tank covering the IGC code requirements using ABS published rules (9). The first set of analyses cover the 8 ABS yield load cases as reproduced in Table 4. As before, the required thicknesses to meet all the yield load cases were determined using Altair's Free-size optimization in OptiStruct. Manufacturability and fabrication considerations dictated the actual partitioning that was used. Once the partitioning was established, gauge optimization was used to establish the required plate thickness in each partition as shown in Figure 7.

The areas of high stresses from the preceding yield load cases were deemed as potential hot spots and were thus further investigated. The finite element meshes in these regions were refined as shown in Figure 7. The hot spot area element size was based on ABS guidelines. Typically, the weld toe areas were where the most critical hot spot stresses develop requiring numerical extrapolation of the finite element results.

7. HIGH CYCLE STRUCTURAL FATIGUE DAMAGE

Per IMO type B tank requirements and ABS guidelines, the cumulative high cycle fatigue damage, D_f is defined as:

$$D_f = D_{f1} + D_{f2}$$

Where

D_{f1}	=	fatigue damage cumulated under load condition
D_{f2}	=	fatigue damage cumulated under normal ballast condition

The cumulative fatigue damage for loading condition i can be calculated as

$$D_{f1} = \frac{1}{6}D_{f1\_12} + \frac{1}{6}D_{f1\_34} + \frac{1}{3}D_{f1\_56} + \frac{1}{3}D_{f1\_78}$$

$D_{f1\text{-}f2}$, $D_{f3\text{-}f4}$, $D_{f5\text{-}f6}$, $D_{f7\text{-}f8}$ are the fatigue damage accumulated due to load case pairs 1 and 2, 3 and 4, 5 and 6, 7 and 8 from the ABS yield loadcases referenced in Table 5. The long term distribution of the stress ranges are assumed to follow a Weibull distribution. The fatigue damage accumulated due to each load pair is derived from the loadcases comprising of the load component coefficients shown in Table 5.

8. THERMAL ANALYSIS AND LOW CYCLE DAMAGE

Thermal stress analysis was carried out to verify the structural integrity of the tank when subjected to thermal stresses during the loading of liquid cargo and initial cooling down period over its life. The conjugate gradient analyses to develop the cool-down strategy is not reported here for brevity, only thermal stress analyses due to cargo loading is summarized here.

As required by code, different filling levels were analyzed to get the cumulative effects of the damage attributed to thermal cycling. The initial state is assumed to start from a heeled out ballast voyage where cargo tank temperature is at -90 C or higher. The cool-down spray-in cycle is then commenced at a predetermined spray-in rate until the tank temperature at the bottom shell of the tank reaches dew point. Cargo loading then commences at a rate of 10% every hour until the tank is filled.

The thermal contractions in the tank during loading introduces significant thermal stresses that have to be carefully assessed and managed. A thermal analysis of the rapid change of temperature from the end of the spray down cycle and the filling step was performed to compute the worst case thermal gradients that could occur at different locations of the tank.

The thermal stresses induced during the filling process were used to calculate the thermal component of the cumulative fatigue damage assuming 1000 loading and unloading cycles over the ship life.

9. TOTAL CUMULATIVE FATIGUE DAMAGE

Total accumulated fatigue damage is assessed as the sum of high cycle and low cycle damage. Total cumulative fatigue damage factor is calculated using:

$$D_T = D_F + \frac{10^3}{N_j}$$

where D_F is the high cycle fatigue damage and N_j is the low cycle damage

Five different filling levels were used to calculate the fatigue damage per ABS rules and the corresponding stresses and damage for the CDTS tank are summarized in Table 6.

Should the total cumulative fatigue damage exceed the allowable limits, the high cycle fatigue damage and/or the low cycle fatigue damage would have to be lowered. While the former damage component would require plate thickness adjustments, the thermally induced fatigue damage could be reduced by lowering the starting temperature after spray-in. This would require the operational protocol to require a higher spray-in rate and/or duration.

10. FRACTURE ANALYSIS

As a Type B tank, a fracture mechanics assessment was done to ensure that the CDTS meets the "leak-before-failure" requirement. Two modes of failure were considered:

a. The initial crack/flaw grows through the wall by fatigue in the event of an overload, to form a stable "through-crack". This can be evaluated by ensuring that crack-length at break-through is below critical crack length or critical crack length corresponds to a crack depth that is higher than the material thickness, ensuring break through before catastrophic rupture
b. An initial surface crack/flaw will not reach critical crack length at the operating stress spectrum over the design life of the ship

These fail-safe scenarios were evaluated to ensure that the CDTS met the proper structural safe-guards outlined above.

Criteria 1

A 0.5 mm deep surface flaw must grow/propagate to exceed plate thickness at an over load condition. We have assumed the overload condition at 75% of the maximum of all the yield load cases. This provides the operators with the warnings associated with LNG leakage well before yield, allowing the ship to reach port where tank inspection and repairs can be made.

Basic governing equation:

$$K_{1c}^2 = M_b^2\left(\frac{\pi\sigma^2 a}{Q}\right)$$

Where

K_{1c} = plane strain fracture toughness, MPa√m (ksi√ins)

M_b	=	Kobayashi magnification factor based upon stress distribution position around crack tip front and crack tip plasticity
σ	=	gross section stress, design stress, MPa (ksi)
Q	=	crack shape factor, including plastic zone correction

Critical crack depth, **a,** is given by:

$\mathbf{a} = (Q/M^2)^* [(K_{1c})^2] / [\pi.^2]$

Using K_{1c} = 55 MPa √m @-162C

A long crack and a circular crack are evaluated and summarized in the Table 7. Criteria 1 is met since the critical crack depth, **a**, is above above the thickest plate used in the design.

Criteria 2

Another aspect of risk mitigation is ensuring crack propagation associated with an initial 5 mm long flaw is controlled and does not reach critical lengths at the operating stress spectrum over the design life of the ship which is assumed at 1 x 10^8 cycles per ABS rules.

Criteria 3

Also, a 0.5 mm deep butt weld flaw or a 1 mm deep fillet weld flaw will not propagate through the thickness at the operating stress spectrum over the design life of the ship which is assumed at 1 x 10^8 cycles per ABS rules.

Crack propagation at the critical location is governed by Paris Equation as follows:

$$\frac{da}{dN} = C(\Delta K)^m \quad \text{for } \Delta K > \Delta K_{th}$$

$$\frac{da}{dN} = 0 \quad \text{for } \Delta K \le \Delta K_{th}$$

Where

da/dN	=	crack propagation rate
C, m	=	Paris constants
ΔK	=	stress intensity factor range
K_{th}	=	threshold value of stress intensity factor range

According published data from test samples, the crack propagation rate for Aluminum grade 5083-O is given

$da/dN = 2.03 \times 10^{-10}$

For crack along surface, an initial 5mm long crack will lead to a maximum propagation length of:

$a_{\max} = 6.74\ mm$

For crack through thickness, an initial 1 mm deep crack will cause the propagation length

$a_{\max} = 1.14\ mm$ (butt welds)

For crack through thickness, a initial 0.5mm deep crack will cause the propagation length

$a_{\max} = 0.55\ mm$ (fillet welds)

Results from criterial 2 & 3 are met and summarized in Table 8

11. LEAKAGE ANALYSIS

IGC code requires the leakage rate of the liquid cargo from the penetrating crack to be analyzed to verify the adequacy of the secondary barrier and drip tray volume. The shape of a crack opening was assumed as an ellipse with crack length and width as major and minor axis diameters. The final size of a crack propagated during 15 days from a penetrating initial crack was used to assess the leakage rate of a liquid cargo.

The leakage rate of the liquid cargo through the crack opening, W in $mm^3/\sec$, per ABS guidelines was determined by the following equation:

$$W = \left(\frac{\pi}{4\mu}\right)\left(\frac{P}{t}\right)\left(\frac{a^3b^3}{a^2+b^2}\right)$$

μ	=	viscosity of the liquid cargo, in $kgf\text{-}\sec/mm^2$
P	=	internal pressure, in kgf/mm^2
t	=	plate thickness, in mm
a	=	a half of the crack length, in mm
b	=	a half of the crack width, in mm

$W_1 = 2.83 \times 10^4\ mm^3/\sec$ (initial)

$W_2 = 3.00 \times 10^4\ mm^3/\sec$ (final)

An appropriately size spill tray was then packaged in to contain the worst case leakage.

12. CONCLUSION

This paper provides an overview of the structural design process of an LNG containment system using the CDTS and a quantitative means of assessing space and weight requirements of different CCSs. While spherical and membrane CCS on LNG ships are proven and both have excellent deployment records, advances in simulation methodologies and manufacturing methods allow for improved design trade-offs to be developed.

The CDTS takes the best features of the existing containment systems like high volumetric efficiency of the membrane, the structural advantages of a spherical tank, and the robustness of the SPB system to offer a system with the highest spatial efficiency, a competitive lightship weight, and ease of fabrication, installation, and best operational lessons learned from existing LNG carriers.

13. ACKNOWLEDGEMENTS

The authors would like to acknowledge with thanks the support of Altair Engineering The application of Altair's advanced CAE tools have enabled us to take the CDTS containment system closer to commercial deployment.

14. REFERENCES

1. LAMB, T, and KOTINIS, M, 'A Set-based Ship Design Synthesis System,'*(2003) IMDC 2003*, Athens, Greece

2. LAMB, T, 'The Effects of Intended Trade Route on the Optimum Size of LNG Tankers,'*(1974) LNG-4 Conference, Algeria*, June 27, 1970

3. LAMB, T, and RAMOO, R, "The Application of a New Tank Containment System to ULTRA-Large LNG Carriers', *OTC* 2009

4. LAMB, T, and RAMOO, R., 'A New Concept for CNG Carriers and Floating CNG/Oil Processing and Storage Offshore Platforms' *CNG Forum*, London 2009

5. LAMB, T, and RAMOO, R, "The Application of a New Tank Containment System to ULTRA-Large LNG Carriers', *OTC* 2009

6. LAMB, T, and RAMOO, R., 'A New Concept for CNG Carriers and Floating CNG/Oil Processing and Storage Offshore Platforms' *CNG Forum*, London 2009

7. LAMB, T, and RAMOO, R, 'A New Tank Containment System for Floating LNG and CNG Processing and Storage Offshore Platforms,' *OTC* 2010

8. GRIN, R., and van de VOORDE, E, 'Weather-Related Economics of Natural Gas Transport for Two Propulsion Plant Configurations,' *RINA Conference, Design and Operation of Gas Carriers*, 2004

9. ABS, 'Guide for Building and Classing, Liquefied Gas Carriers with Independent Tanks', 2010

10. Altair HyperWorks 14.0 Online Help

11. Fracture Toughness and Fatigue properties of 5083-O Plate and 5183 Welds for Liquefied Natural Gas Application, J.G. Kaufman and R.A. Kelsey

15. AUTHORS BIOGRAPHY

Mr. Ramoo, Vice President of Engineering at Altair Product Design has over 30 years' experience in the use of computational methods for design ideation and Advanced Structural Analysis. Mr. Ramoo graduated with a Diploma in Marine Engineering from the Singapore Polytechnic and holds a MSc. in Mechanical Engineering from The Ohio State University. He has broad experience in using advanced CAE methods in structural design of land vehicles and marine structures.

Mr. Eng, Engineering Manager at Altair Engineering has over 10 years' experience in the use of computational methods in structural analysis. Mr. Eng graduated with a MSc. in Engineering Science and Ocean Engineering from The National Taiwan University in 2005. His experience is in heavy industry and ship structure design using advanced CAE methods.

Mr. Wu is a Project Engineer at Altair Engineering. Mr. Wu graduated with a MSc. in Engineering Science and Ocean Engineering from The National Taiwan University on 2015. His project experience has been in ship structure design and strength analysis.

Mr. Lamb has over 60 years' experience in the design and construction of many ships of different types and spent his last 11 active years teaching and performing research at University of Michigan, retiring in 2006. He served a five-year apprenticeship at H.M. Dockyard Rosyth and graduated from Durham University in 1958 with his B.Sc. in Naval Architecture. He is a Fellow of RINA, and SNAME, and was the Editor of and 5-chapter contributor to the 2003 edition of the SNAMESHIP DESIGN AND CONSTRUCTION book as well as being the SNAME Ship Design Committee Chairman for many years. He is a Professional Engineer in the U.S., Chartered Engineer in UK and Eur Ing.

Table 1:
Comparison of Containments Systems

Ship Type	Hold Volumetric Efficiency hv	Lightship Weight Efficiency hw (relative to membrane)	Combined "Efficiency" (Utilization Factor Uf)	Cost Index	GT Index
Membrane	48%	100%	48%	1.1	1.0
Spherical	45%	88%	40%	1.0	1.24
SPB	40%	77%	31%	1.3	1.09
CDTS	63%	95%	60%	0.9	0.97

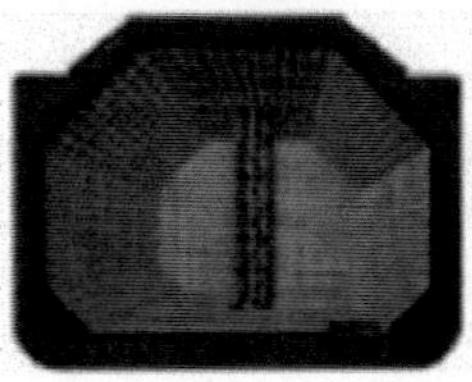

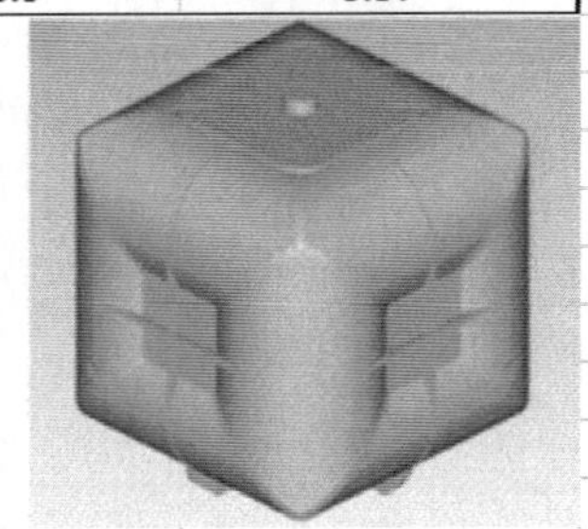

Table 2

CCS Type	CCS Weight Index	Lightship Weight (170K)	Capacity/ Gross Tonnage
Membrane	0.85	36907	1.54
Spherical	1	42141	1.25
SPB	3.2	47898	1.42
CDTS	1.9	38652	1.60

Table 3:
MSDSS Results

		MSDSS DESIGN SUMMARY				Published	
CHARACTERISTIC		170k Spherical	170 k Membrane	170k SPB	170 k CDTS	Energy Horizon Moss	Creole Spirit Membrane
LBP	m	285	279	290	265	286.5	282.9
Beam	m	51.4	48	50.2	48	52	46.4
DEPTH MAIN DECK	m	27.8	27	26	26	28	26.5
DEPTH COVER	m	20	11	0	15	20	10.1
DRAFT	m	11.5	11.5	11.5	11.5	11.676	12.5
SPEED	Knots	19.5	19.5	19.5	19.5	21.8	
BLOCK COEFFICIENT		0.701	0.748	0.728	0.804		
DISPLACEMENT	t	120150	115330	125401	116943		
LIGHTSHIP	t	42141	36907	47898	38652		
DHP	KW	27963	29304	29115	35336	29890	36400
DEADWEIGHT		81438	81426	81400	81437	87257	83400
Gross Tonnage		136520	110175	119803	106407	141136	113263
CARGO CAPACITY	C M	170053	170099	170251	170002	177441	173400
Capacity/Gross Tonnage		1.25	1.54	1.42	1.60	1.26	1.53

Table 4: ABS Standard Design load cases for Yielding and Buckling Strength Assessment
(Ref: ABS, 'Guide for Building and Classing, Liquefied Gas Carriers with Independent Tanks')

	LC1	*LC2*	*LC3*	*LC4*	*LC5*	*LC6*	*LC7*	*LC8*
Dominant Load Parameter	maximum sagging BM	maximum hogging BM	maximum internal pressure	maximum external pressure	maximum internal pressure	maximum horizontal BM	maximum horizontal BM	maximum internal pressure
Wave Heading (μ)	0° head	0° head	0° head	0° head	90° beam	60° oblique	60° oblique	0° head
Draft (d_m)	d_f	d_f	d_f	d_f	$3/4d_f$	$3/4d_f$	d_f	d_b
External Pressure k_{ce}	–0.50	0.50	–0.50	1.00	1.00	–0.50	1.00	–0.50
Longitudinal Acceleration $k_{c\ell}$	0.00	0.00	0.00	0.60	0.00	–0.40	0.40	0.00
Vertical Acceleration k_{cv}	0.50	–0.50	1.00	–0.80	–0.60	0.60	–0.60	1.00
Transverse Acceleration k_{ct}	N/A	N/A	N/A	N/A	0.80	–0.70	0.70	N/A
Vertical BM k_{cmv}	1.00 sagging	1.00 hogging	0.70 sagging	0.70 hogging	0.30 hogging	0.30 sagging	0.30 hogging	0.70 sagging
Horizontal BM k_{cmh}	N/A	N/A	N/A	N/A	0.225 port (+)	0.75 port (–)	1.00 port (+)	N/A
Frictional Coefficient	N/A	N/A	N/A	N/A	0.10	0.10	N/A	N/A

Table 5: ABS Standard Design Load Cases for Fatigue Strength Assessment (Ref: ABS 'Guide for Building and Classing, Liquefied Gas Carriers with Independent Tanks')

Load Case Pair	*LC1 & LC2*		*LC3 & LC4*		*LC5 & LC6*		*LC7 & LC8*	
Wave Heading (μ)	0° head		0° head		90° beam		60° oblique	
Local Pressure Load Case	*LC1*	*LC2*	*LC3*	*LC4*	*LC5*	*LC6*	*LC7*	*LC8*
Draft (d_m)	d_f	d_f	d_f	d_f	d_f	d_f	d_f	d_f
External Pressure k_{ce}	–0.50	0.50	–1.00	1.00	–1.00	1.00	-1.00	1.00
Longitudinal Acceleration $k_{c\ell}$	0.00	0.00	0.00	0.00	0.00	0.00	0.00	0.00
Vertical Acceleration k_{cv}	0.50	–0.50	1.00	–1.00	0.40	–0.40	0.20	–0.20
Transverse Acceleration k_{ct}	N/A	N/A	N/A	N/A	–0.70	0.70	–0.40	0.40
Vertical BM Load Case	100% Rule Vertical BM Range		50% Rule Vertical BM Range		20% Rule Vertical BM Range		30% Rule Vertical BM Range	
Horizontal BM Load Case	N/A		N/A		30% Rule Horizontal BM Range		50% Rule Horizontal BM Range	

Table 6: Damage Calculation for Hot Spot

	High Cycle Damage	Thermal Damage	Total Damage
Hot Spot 1	1.59E-03	0.12	0.12159
Hot Spot 2	4.00E-04	0.15	0.1504
Hot Spot 3	1.97E-04	0.001	0.001197

Table 7: Critical length of different stress range

			stress (Mpa)			
		1.09884	68	48	40	20
Plate Thickness t (ins)	**Fracture toughness K_c ksi $\sqrt{(ins)}$**	**Fracture ToughnessK_c MPa $\sqrt{(m)}$**	**Critical Length a_c (m)**	**Critical Length a_c (m)**	**Critical Length $a_{c\ (m)}$**	**Critical Length $a_{c\ (m)}$**
2	124	136.26	1.28	2.56	3.69	14.77
4	103	113.18	0.88	1.77	2.55	10.19
6	85	93.40	0.60	1.21	1.74	6.94
For Through Thickness Crack	use K1c	55	0.21	0.42	0.60	2.41

Table 8: Long crack calculation

	Operating Stress Load Cases (Max. of Operating Stress Load Cases)		Overload Stress Load Cases (75% Max. Yield Load Case)	
Q	1.25	2.4	1.25	2.4
M	1.1	1.1	1.1	1.1
$K1c$	55	55	55	55
σ	48	48	73.5	73.5
a(mm)	432	829	184	354

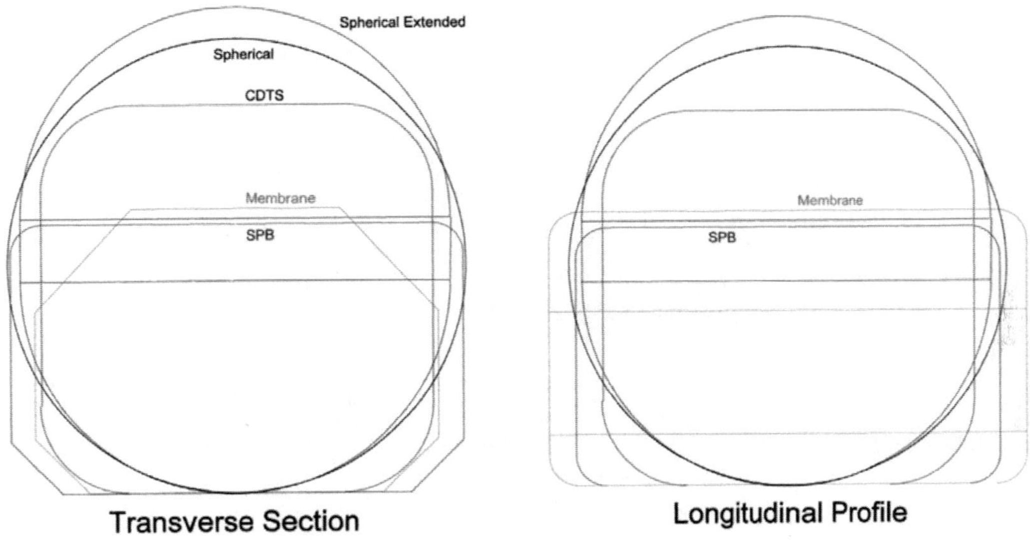

Figure 1: Four containment system shape comparison

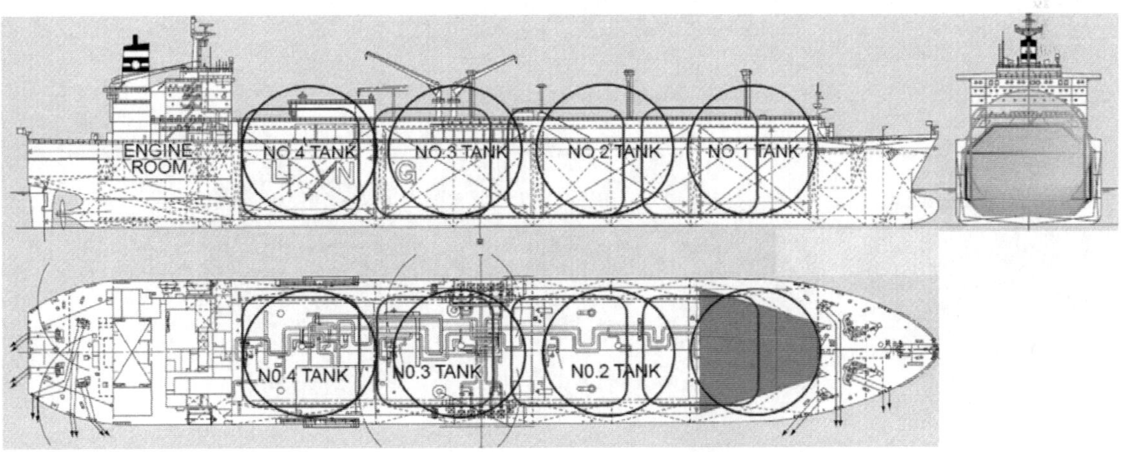

Figure 2: Typical layout of membrane ship

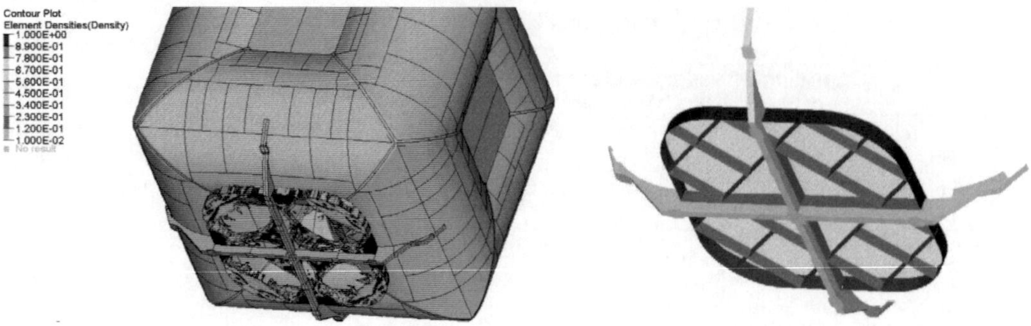

Figure 3: Topology optimization and ribs structures of base endcap structure

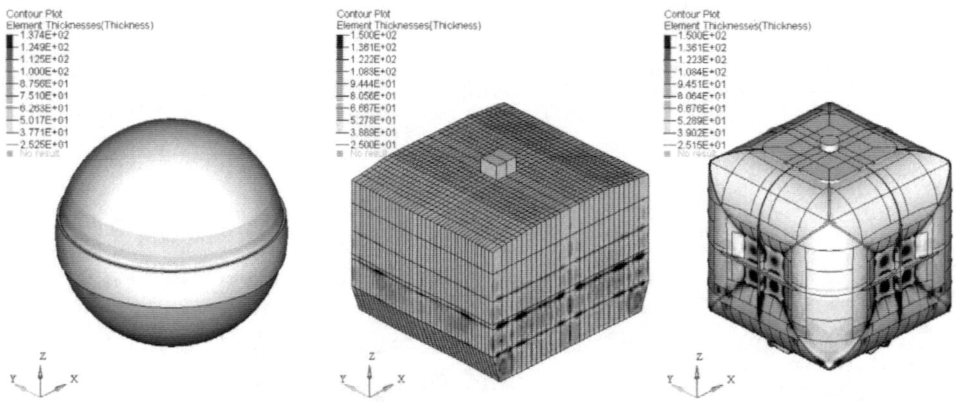

Figure 4: Free-size Optimization of 3 types of Containment Systems (mm)
(From Left: SPHERICAL, SPB, CDTS)

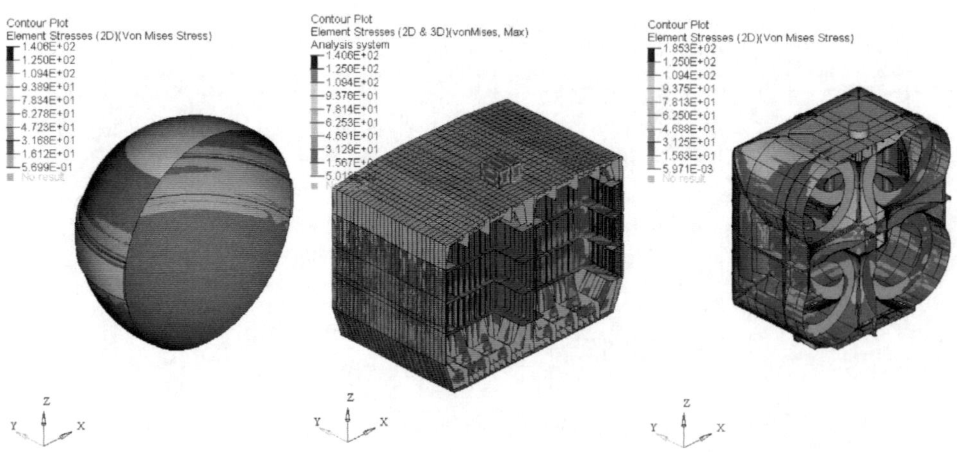

Figure 5: Stress distribution of MOSS, SPB and CDTS tank (MPa)

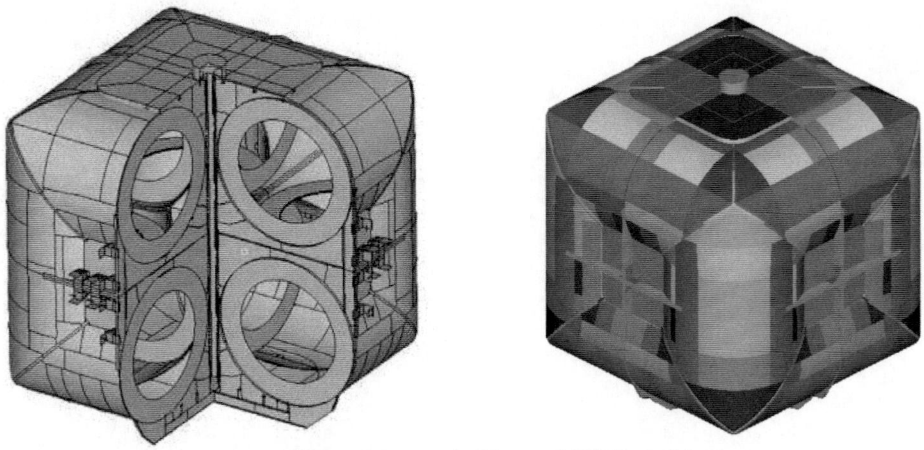

Figure 6: Isometric Views of CDTS model

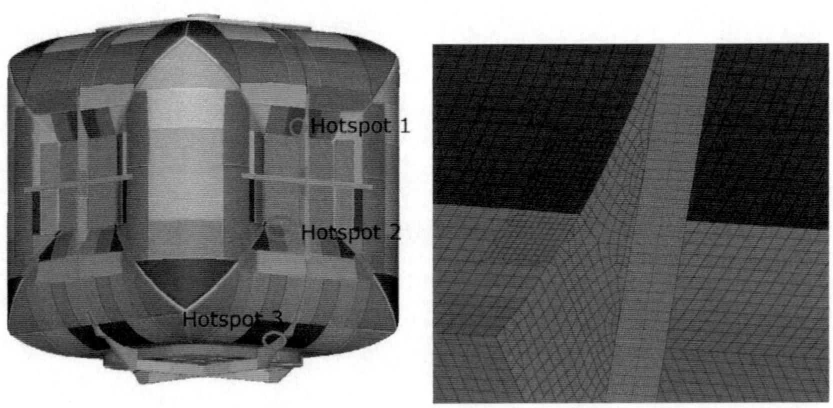

Figure 7: HotSpot location and Element Size at Hotspot 3

DEVELOPMENTS IN FRICTION STIR WELDING FOR LNG SHIPPING

S Cater FRIN MEI AWeldI, **J Martin** CEng SenMWeldI, TWI Technology Centre (Yorkshire), UK

SUMMARY

The transport of LNG imposes demanding requirements on the construction of vessels, requiring materials and high integrity welding techniques with excellent fatigue life and high toughness at cryogenic temperatures. Friction stir welding (FSW), an innovative solid state joining technique, produces welds with a forged microstructure rather than the cast microstructure of a traditional fusion weld and thus can provide very high strength welds. The process can weld all grades of aluminium alloys in thicknesses considerably in excess of 50mm and is now being introduced for the construction of LNG carriers.

Recent developments in the FSW process, for example the ability to generate a reinforced fillet as part of the welding process, have further expanded the capabilities of the technique and extended its suitability for the manufacture of ships and pressure vessels from thick section aluminium. Research is also underway to extend the capability of the FSW process to steel for marine construction applications, with the recent EU project HILDA demonstrating the suitability of the FSW technique for the fabrication of hull and deck panels in 6 and 8mm thick DH36 carbon steel.

1. INTRODUCTION

Though steel is the dominant construction material for larger ships, providing for the manufacture of strong, stiff hulls at moderate cost, there is a growing usage of aluminium alloys within the specialist shipbuilding sector. Though aluminium alloys are more expensive than steels, material costs are typically less than 3% of the manufacturing cost of a ship and thus savings arising from light-weighting and corrosion resistance can offset the initial cost premium. Aluminium alloys' lower mass makes them very attractive as a means of reducing structural weight, increasing a vessel's speed and payload capacity or reducing fuel consumption. In addition, aluminium alloys' excellent corrosion resistance can aid offset maintenance costs and thus its initial higher purchase cost with respect to steel. The most widespread aluminium alloys for shipbuilding are AA5083, a non-heat treatable, high strength grade frequently used for major structural elements, and the heat treatable 6082 alloy widely used for panelling, decking and superstructures. In certain specialist applications aluminium alloys can outperform steel: a typical example of this is in the construction of LNG carriers where AA5083 exhibits better toughness properties at cryogenic temperatures than steel.

Aluminium alloys have often been perceived as being more difficult to weld than steel, requiring more complex welding procedures and more highly skilled welders. However, the invention of friction stir welding (FSW) in 1991 by TWI[1] greatly simplified the process of welding aluminium and cuts costs significantly. Friction stir welding is a simple, solid state, mechanical welding process that is easily automated: it thus eliminates the requirement for skilled welders and significantly increases productivity when fabricating structures from wrought plate or extrusions. The basic process of friction stir welding is shown in Figure 1.

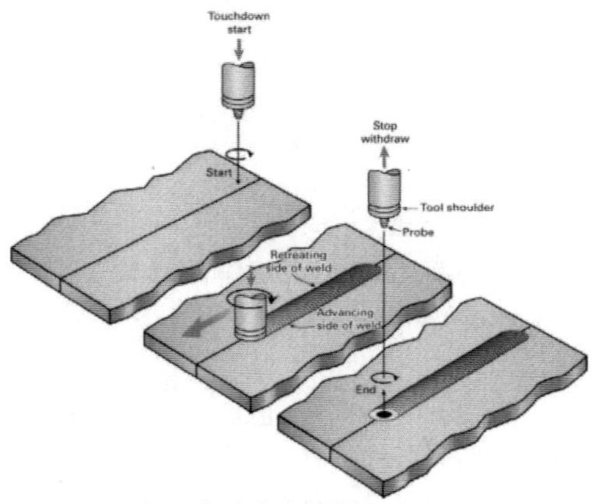

Figure 1: The principle of FSW.

To make a butt weld between two metal plates, they are first butted together and clamped rigidly in place. A specially designed rotating tool is lowered onto the surface of the plates and begins to soften the metal by the generation of frictional heating. As the metal softens, the tool embeds to the full depth of the plates and is then traversed along the desired joint line, mechanically stirring the two plates together. At the end of the weld, the tool is extracted. The metal being welded is never molten, only plasticised. Friction stir welding has been the subject of rapid development and application across the world since its invention, and is now a well-established joining technique for aluminium alloys.

Benefits of the friction stir welding process include:

- Solid phase process.
- Single pass process (~0.5mm to >75mm per pass in Al).
- Fully mechanised process, therefore reducing the need for skilled welders.

[1] Thomas W M, Nicholas E D et al: 'Improvements Relating to Friction Welding', European Patent Specification 0 615 480 B1, December 1991

Figure 2: The use of friction stir welded aluminium for ship construction

- Some tolerance to poor pre-weld edge profiling or cleaning.
- No shielding gas or filler wire required for most materials.
- Low distortion and shrinkage in most cases.
- 1-D, 2-D or 3-D, any position.
- Excellent mechanical properties.
- Physical properties (e.g. conductivity) = to parent material.
- No welding fume or spatter hazards.
- No UV or electromagnetic radiation hazards.
- Modest energy consumption.
- Generally quiet process.
- Can weld underwater

2. EARLY CONSTRUCTION IN ALUMINIUM USING FSW

The automated, high quality, high productivity, low cost nature of the friction stir welding process saw its rapid adoption in many shipyards, particularly for the prefabrication of panels for decking and superstructure. Figure 2 shows examples of the use of FSW for pre-fabrication of panels and superstructure elements.

Early applications in friction stir welding for marine construction centred primarily on the joining of plate and extrusions using butt welds in order to fabricate larger panels. However, many applications require the use of a reinforced fillet weld in order to comply with design codes aimed at reducing stress concentrations in fabricated corners. Though it might seem that a solid state welding technique would be unable to produce a fillet weld as there is no molten weld pool to which material can be added, this is not the case. TWI has developed several variant techniques based upon stationary shoulder friction stir welding (SSFSW) which allows for the creation of fillet welds or the joining of materials of different thicknesses.

3. FILLET WELDS AND DISSIMILAR THICKNESS WELDS

One technique for creating a fillet weld is AdStir corner SSFSW, in which the tool's rotating probe is surrounded by a stationary shoulder, the leading and trailing edges of which are of different shape. (Figure 3). The leading edge has a small aperture into which a filler wire is drawn by the rotating probe as this plasticises and welds the two plates together. The filler wire is also plasticised, creating a small excess of metal which is then forged to a fillet by the trailing edge of the stationary shoulder which is appropriately shaped.

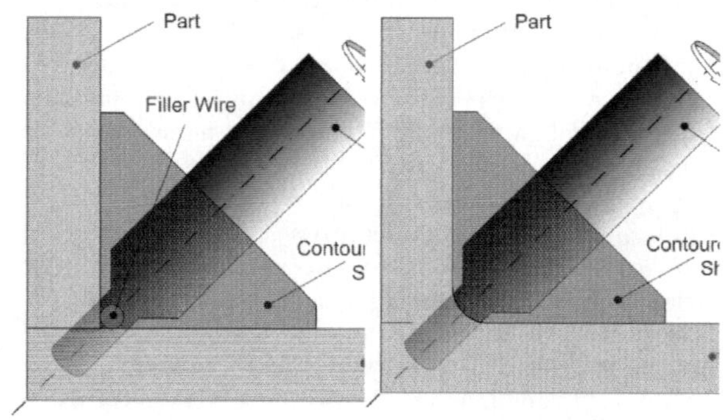

Into the weld view Away from the weld view

Figure 3: Schematic of AdStir corner SSFSW technique

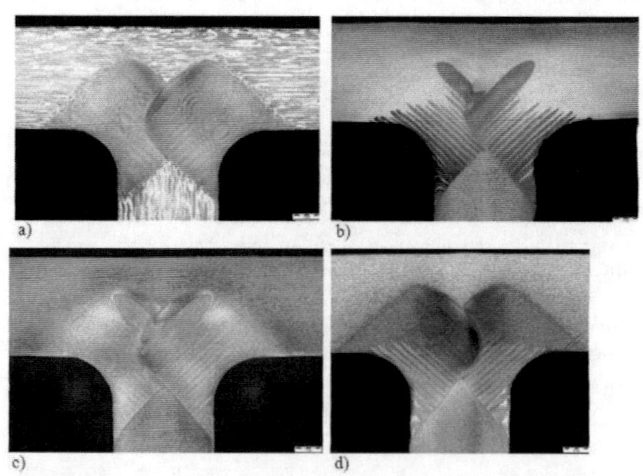

Figure 4:
a) AA6082-T6 base and leg with a AA6082-T6 filler;
b) AA5083-O base and leg with a AA6082-T6 filler;
c) AA5083-O base and leg with a AA5356-T6 filler; and
d) AA2014-T6 base and 7075-T6 leg with a AA6082-T6 filler

Figure 4 shows several macrographs of polished and etched T sections fabricated by AdStir corner SSFSW from 8mm thick plates. The plates can be of different thickness and of different aluminium grades. The filler wire can also be of a different aluminium grade, and may even contain particulates to create a localised metal matrix composite in the weld zone if required. A welding pass made from both sides of the upstand can be used to create a full penetration weld.

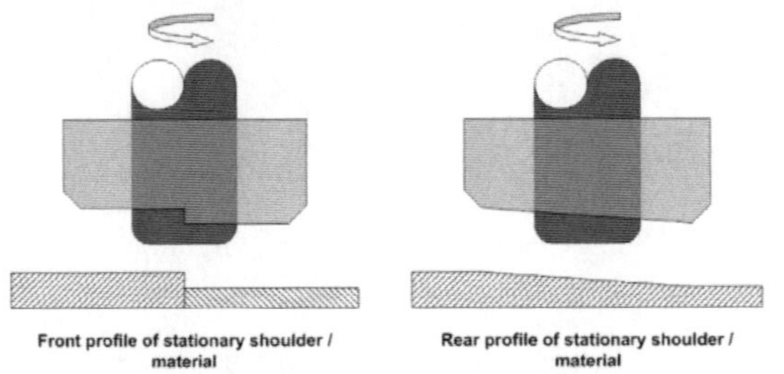

Figure 5: Profiles of stationary shoulder used for dissimilar thickness welds

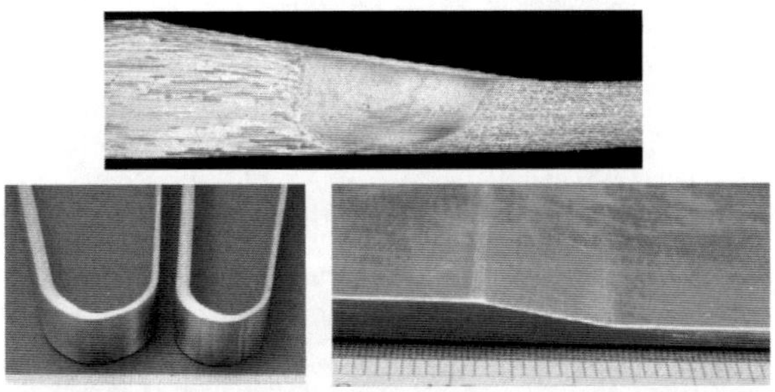

Figure 6: Example of a joint made between 4mm and 2 mm thick AA6082 sheet. The macrograph (upper) shows the weld to be defect free and the root and face bend test samples (lower left) attest to its mechanical integrity. The smooth transition between the two very different thicknesses of material is shown lower right.

The ability to have different profiles on the leading and trailing edges of a stationary shoulder allows for the forging and forming of the plasticised material into different shapes. This can be utilised, for example, to weld different thicknesses of material. Figure 5 shows the basic principle of the forming process, highlighting the differently shaped leading and trailing edges of the stationary shoulder, and Figure 6 shows a joint made between 4mm and 2mm thick sheets of AA6062. The welded joint has a smooth surface finish at the transition between the two sheets of material and the weld itself is defect free, passing face and root bend tests.

4. CONSTRUCTION IN THICK SECTION ALUMINIUM ALLOYS

The majority of early marine construction applications using FSW to join aluminium components were in the thickness range of 2-8mm. However, there are many applications for which joining of thick section aluminium components is required, for example bulkheads and LNG tanks. Using conventional fusion welding techniques for welding thick aluminium alloys can be challenging, as typically:

- A multi-pass technique is required, leading to low productivity;
- Component edge preparation and a shielding gas are necessary;
- The process is difficult to control;
- Preheat is required;
- Welds are prone to porosity and cracking;
- Welding can lead to component distortion;
- Some aerospace alloys in the 2xxx and 7xxx series are difficult, or impossible, to fusion weld.

TWI has demonstrated that by using FSW it is possible to weld thick section aluminium alloys and that these welds possess the excellent mechanical and fatigue properties required for safe operation at cryogenic temperatures. The process has been adopted for the construction of LNG tanks by IHI Corporation in Japan with certification by ClassNK[2]. Figure 7 shows a section through IHI's Self Supporting Prismatic Type B (SPB) LNG tank. This is an insulated aluminium tank contained within a steel hull.

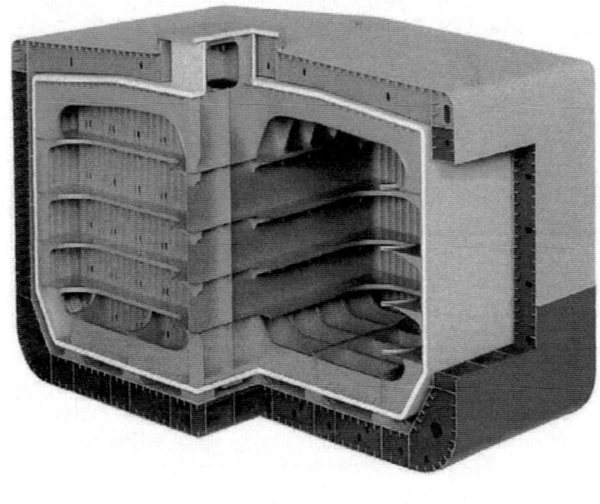

Figure 7: IHI Corporation's IHI-SPB LNG tank.

[2] IHI Engineering Review, Vol 49, No 1, 201,6 pp 10-13

Figure 8a: TWI PowerStir™ FSW machine and Figure 8b: right, a weld being made in thick section aluminium plate using the simultaneous dual welding head process

Aluminium alloy LNG tanks are typically manufactured from AA5083 plate up to 25 mm thick. It is possible to friction stir weld aluminium alloys of considerably greater thickness than 25mm, and this offers considerable potential for fabricating even larger structures. However, as the thickness of the material to be welded increases several problems arise, including:

- Short tool life due to high welding forces;
- Very slow welding traverse rates;
- High heat input leading to component distortion.

Typically, to weld thicker plate, a Weld-Flip-Weld approach is used. This consists of using a FSW tool with half thickness component penetration. After the first weld has been made, it is normally necessary to remove any flash (excess plasticised aluminium) from the weld surface before releasing the component from the fixture. Some plate distortion can occur, so once the partially fabricated plate has been lifted and turned, it must be placed back in the fixture and forced into position prior to completing the second weld. These additional operations (component cleaning, pilot holes, handling, and refixturing), especially when fabricating large and heavy components, can be a very significant portion of the total time required to fabricate the component.

In response to this, TWI has been developing several new FSW techniques to minimise cost, production cycle time, and weld heat input, making joining thick section aluminium alloys both technically and commercially more advantageous. All these techniques utilise the dual head welding capabilities of the PowerStir FSW machine (Figure 8a) at the TWI Technology Centre in Yorkshire. This multi-functional R&D machine has opposed, synchronised spindles. Although there are many FSW machines with opposed spindles, it is believed that this is the only machine that can run tooling capable of at least 50mm penetration on each head. The upper and lower spindles housing the two FSW tools, and also the slotted fixture used for welding in a simultaneous configuration, are shown in Figure 8b.

Initial trials, when simultaneously welding from either side of 50mm thickness aluminium alloy plates, have shown that the welding speed can be significantly increased when compared with the Weld–Flip–Weld joining technique. This is explained by the additional energy from the two tools and the slotted fixture not conducting the heat away. This local heat produces a larger, softer plasticised region around the FSW tool, providing less resistance to the passage of the tool through the material for the same tool traverse rate. A typical weld section is shown in Figure 9.

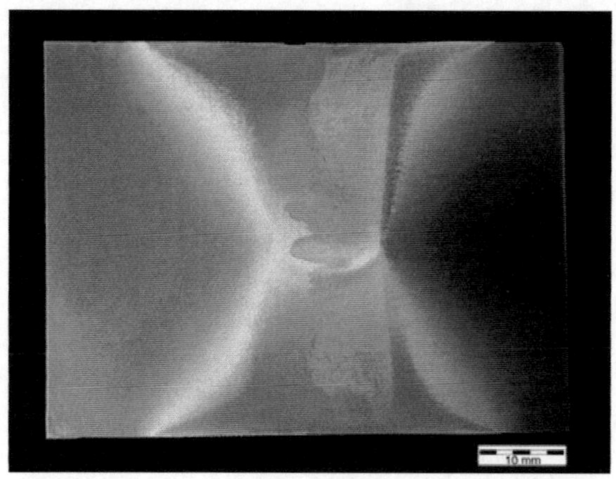

Figure 9: A metallographic section through a simultaneous dual headed friction stir weld in thick aluminium plate.

A doubling of the welding speed has been achieved for similar traverse welding forces experienced by the FSW tool. The effect on the total process cycle time using simultaneous welding is very significant, as shown in Figure 10. Not only is the welding time reduced, but secondary operations such as handling dressing and refixturing of a potentially distorted component, are eliminated. However these benefits have to be considered against the increased capital cost of ownership of a more capable FSW machine.

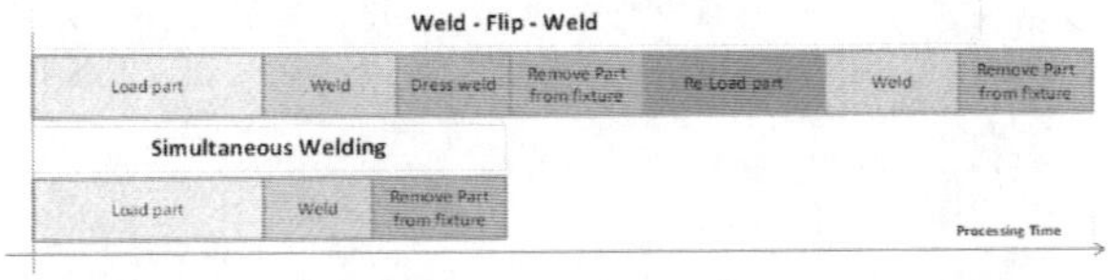

Figure 10: Comparison of cycle times for welding thick section aluminium.

TWI is developing other techniques for welding thick section components in a single setup. The first of these techniques is called Supported Bobbin FSW. TWI has developed the floating bobbin FSW technique which has many advantages over conventional FSW however, as the component thickness increases; it becomes increasingly difficult to control the position of the bottom bobbin shoulder. Figure 11 shows a macrograph taken through a weld made in 40mm thickness AA6082-T6, which shows that the upper and lower weld surfaces are not parallel. There is a narrowing of the section on the weld advancing side (right) and a thickening on the retreating side. This phenomenon was termed 'wedging'. The probe connecting the upper and lower shoulders of the bobbin is therefore rotating in an abnormal manner during welding, which leads to premature tool failure

Figure 11: A macrograph taken from a bobbin weld made in 40mm thickness AA6082-T6

In the Supported Bobbin technique, the bottom shoulder is radially restrained by a bearing in the under head as shown in Figure 12. An example of a weld made using this technique is shown in Figure 13. This weld was made in 50mm thick AA5083-O, produced using a welding speed of 85mm/min, without tool failure, and the upper and lower weld surfaces were nominally parallel. A comparison on the potential benefits of the supported bobbin FSW technique to fusion welding 50mm thickness AA5083-O is given in Table 1.

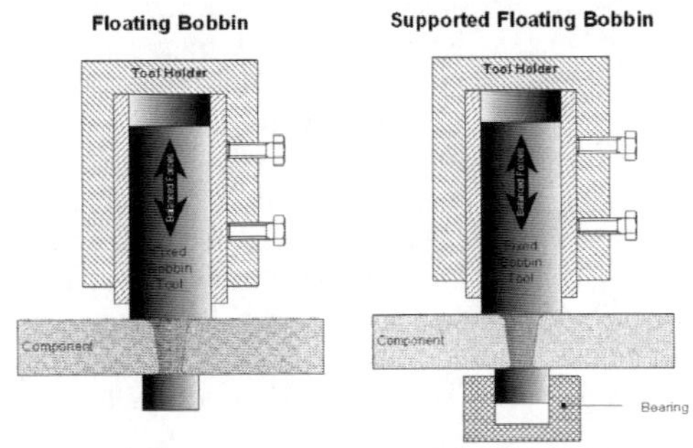

Figure 12: Schematic drawing of Floating Bobbin FSW and Supported Bobbin FSW

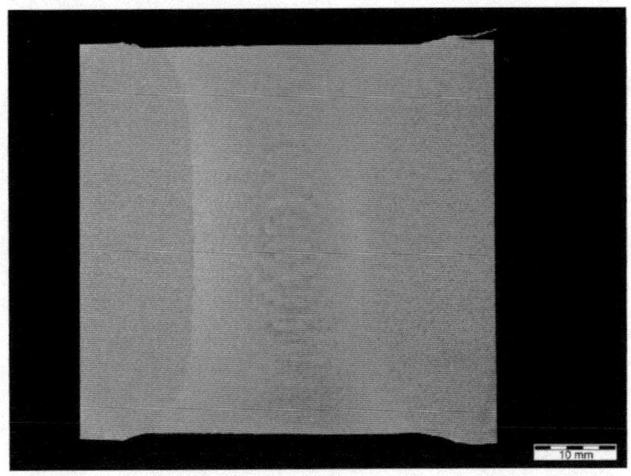

Figure 13: Macrograph taken from a supported bobbin weld in 50mm thickness AA5083-O

Table 1: Comparison between fusion and FSW welding a one metre long component 50mm thickness AA5083-O

Fusion	**FSW**
Recommended double sided weld (Pre heating)	One setup welding technique
V-Prep on both sides (~two hours)	None
Consumable wire diameter 1.2mm; Weight around 10Kg	None
25 passes each side (Total of 50 passes for both sides)	1 Pass
Flip plate over after each pass (To reduce distortion)	None
Average weld speed 1m/min	Weld speed 85mm/min
50 min welding time	12 min welding time
Total of six hours including prep and handling	Total of one hour including handling
Weld UTS around 95% of parent material	Weld parent strength
Expected distortion in transverse plane, around 10 degree	Expected distortion in transverse plane, less than 1 degree

Further trials were undertaken to assess the performance of smaller sized bobbin tools. Using conventional floating bobbin tools to join 25mm thickness AA5083-O, the maximum welding speed which could be realised was 100mm/min without 'wedging'. Using a bobbin tool design featuring a smaller probe and shoulders, it was possible to increase the tool rotation speed, and also welding speed to 180mm/min, using the supported bobbin technique. The calculated spindle energy required to make these welds reduced from 4.7kJ/mm for the larger tool to 2.5kJ/mm for the supported smaller tool.

The benefits of using smaller tooling were demonstrated, as welds could be made at higher traverse speeds. Supporting the lower shoulder assisted in reducing tool bending however as the probe size reduced it had insufficient tensile strength to contain the plasticised materials leading to probe failure. To try to reduce / eliminate the tensile stress on the probe a new FSW variant was designed called Sliding Bobbin. In this technique, both upper and lower shoulders are driven. The probe is held in the upper shoulder but the lower end of the probe is located and held radially in the bottom shoulder but allowed to slide longitudinally. Only a limited number of welds have been made using this technique but it shows great promise. Similar results have been obtained to supported bobbin using the 16mm diameter probe, smaller diameter probes have yet to be assessed. Although this technique requires two driven heads, sliding bobbin friction stir welding has the following potential benefits over simultaneous double sided welding:

- Use of small diameter probe/shoulders and therefore increased welding speed;
- Lower heat input;
- More even heat input;
- No double processed area in the weld;
- Less complex welding zones simplifying NDT assessment

The third new technique is called Supported FSW, and is based on the dual-rotation FSW technique. In these trials, the shoulders were kept stationary, and only the probe rotated. It was expected that this technique, similar to the supported bobbin technique, would:

- Increase the weld depth possible by a single weld;
- Use small diameter probes/shoulders, and therefore increase welding speeds;
- Eliminate lack of penetration defects.

With this technique, as heat is generated only by the probe, a lower weld heat input is expected. This may be beneficial when joining aluminium alloys, which obtain their strength by solution heat treated and aging, as less thermal damage to the material properties would occur. A limited number of welds have been made using this technique, and it has been demonstrated that it is possible to make sound welds as shown in Figure 14.

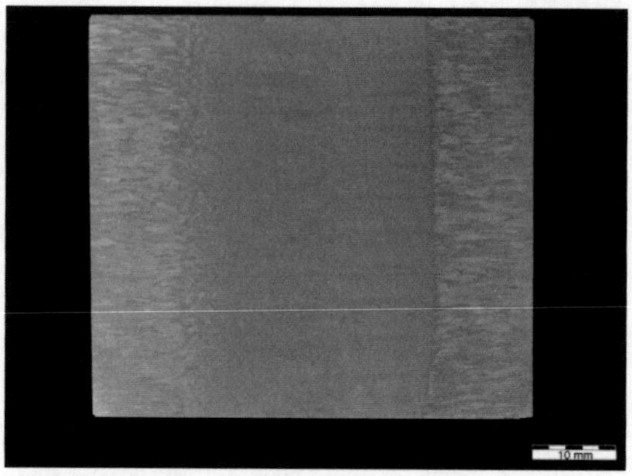

Figure 14: A welding made using Supported FSW in 50mm thickness AA6082-T6

5. CONCLUSION

Recent innovations at TWI, using the unique twin headed capabilities of the PowerStir FSW machine, have demonstrated several techniques which have the potential to minimise production cycle times, making joining of thick section aluminium alloys both technically and commercially advantageous. These techniques need further development to understand more fully the extents of the process window, weld properties, and limits on tool geometry; however, this work has demonstrated the potential to:

- Reduce component processing times through shorter weld times and minimal part handling;
- Improve weld quality through use of reduced and uniform heat input;
- Increase tool life leading to reduced tooling cost;
- Reduce distortion.

6. AUTHOR'S BIOGRAPHIES

Jonathan Martin CEng Sen MWeldI is Section Manager and Technical Lead in The Welding Institute's Friction and Forge Processes Department . He is responsible for overseeing the department's work on process development and technical innovation, a role he has performed since 2005.

Stephen Cater FRIN MEI AWeldI is Principal Project Leader in TWI's Friction and Forge Processes Department. He is responsible for the friction stir welding of thick section aluminium alloys and the development of the FSW process in steel, winning the Richard Weck Award for the latter in 2012. Stephen also lead TWI's input to the 3 year EU sponsored projects HILDA (High Integrity Low Distortion Assembly) and MOSAIC (Materials On-board, Steel Advancements and Integrated Composites), both of which sought to assess the feasibility and benefits to be gained from the introduction of friction stir welding to the shipbuilding sector.

AUTHOR'S CONTACT DETAILS

Akira Nakashima
Yokohama National University
nakashima.akira@jgc.co.jp

Alan Crowle
CB&I
acrowle@CBI.com

Anders Bjoern
Kosan Crisplant
abj@kosancrisplant.com

Andrew Clifton
SIGTTO
manager@sigtto.org

Carlos Guerrero
Bureau Veritas Marine & Offshore Division
carlos.guerrero@bureauveritas.com

Darko Bobić
University of Belgrade
dbobic@gmail.com

Geanni Sicuro
Trelleborg
kwilole@blue-comms.com

Henry Chung
Lloyd's Register
henry.chung@lr.org

John R Maguire
Independent Consultant
john.maguire@sky.com

Keith W Hutchinson
Babcock International
Keith.W.Hutchinson@babcockinternational.com

Nikola Vladimir
University of Zagreb
nikola.vladimir@fsb.hr

Regu Ramoo
ALTAIR Engineering
rrr@altairpd.com

Stephen Cater
TWI Technology Centre
Stephen.Cater@twi.co.uk

Thierry Clément
GTT
tclement@gtt.fr

Thomas Lamb
University of Michigan
nalamb@umich.edu